THE KANSAS CITY UFO FLAPS

THE KANSAS CITY UFO FLAPS

By Margie Kay

Published by UnX News Media
PO Box 1166
Independence, Missouri 64051

ISBN: 978-0-9988558-1-3

Cover art: Alien mother-ship UFO nearing Earth, with rising moon © 3000ad– Fotolia.com
stars in space or night sky © clearviewstock– Fotolia.com

UNXMEDIA

PUBLISHING

Dedicated to

Jean Walker- for her support and encouragement
Joe Palermo- for his unique understanding of these events
Maria Christine, my daughter- for her insight and advice
Debbie Ziegelmeyer- for her outstanding leadership and dedication
and Stanton Friedman- for his inspiration and determination

Special thanks to the following persons who assisted with investigations:

Debbie Ziegelmeyer, State Director for Missouri MUFON

Race Hobbs, Field Investigator in Arkansas

Stan Seba, State Director for Kansas MUFON

Tamie Lynne, my office assistant and right hand woman

And other Missouri Investigators who assisted with cases

Also special thanks to the National UFO Reporting Center and International MUFON

Table of Contents

UFO Hotspot: Kansas City

Kansas City, Missouri and the surrounding area has a long history of UFO sightings and strange phenomena, going back to the airship sightings of 1896—1915. The first airship was spotted over Sacramento, California in November of 1896. The airships were large and looked similar to dirigibles which were not invented at that time. Many had portholes, or had people hanging out of windows, or over railings. Some of the passengers spoke to witnesses. These strange encounters occurred across the country, with several sightings in Kansas City. In one strange case, two "people" were seen to be floating in the air as if they were swimming through water. Most of the sightings were in 1897, but several occurred in later years.

UFO sighting flaps in the greater Kansas City area occurred in 1989—1989, and again in 2011-2014. A "Flap" is a large number of UFO reports in a given time frame.

However, the number of reports have increased in recent years for inexplicable reasons. Kansas City had the highest number of UFO sighting reports submitted to International MUFON during the month of October, 2011. A total of 93 UFO sighting reports were submitted to Missouri MUFON and directly to me during this period with most of the sightings occurring in the greater Kansas City, Missouri area. These areas including the suburbs of Lee's Summit, Raytown, Liberty, and Independence. The average number of reports for a 30-day period in the Kansas City area prior to this time period was normally three or four, so this was a significant increase. It is notable that sighting reports increased steadily in the area since April of 2011, and continued to be high in number until the latter months of 2012, when reports dropped off slightly only to increase once again in 2013 and 2014.

What I find unusual is the fact that many of these reports have been close encounters of 500 feet or less, sightings of non-human entities, missing time, and abductions rather than the typical "lights in the sky" scenario that is most common among UFO sightings. As you'll see, some of the witness accounts are nothing short of incredible. Some of the sightings are my own and other investigator's which occurred while we were on investigation sites or going about our daily routines.

Many people have asked the question "Why Kansas City?" This has crossed my mind many times and I've been searching for answers. Kansas City has some unique features – a large underground natural cave system, The Missouri River and Little Blue River along with Longview Lake, Blue Springs Lake, Lake Jacomo, and other smaller lakes and ponds. Perhaps if UFO occupants need to hide somewhere they can find a place here.

Kansas City is a fifteen-county metropolitan area with a population of 2,343,000 spanning the Kansas and Missouri borders. The Missouri river divides the area into what locals call "North of the River" and "South of the River." The Kansas River is situated to the west, where it joins the Missouri River. Do UFOs need water or need somewhere to hide? If so, they can get it in Kansas City.

There is a fault system under the city of Raytown, and a recently discovered fault line branch from the New Madrid fault that reaches underneath the tarmac at the Richards-Gebaur AFB near Grandview, just south of Kansas City. Many UFO investigators have noted that balls of light and UFOs are often seen just prior to earthquakes. Perhaps some of these preceded the

B-2A returns from a mission as part of Joint Task Force Odyssey Dawn

By U.S. Air Force photo by Senior Airman Kenny Holston - http://www.flickr.com/photos/usairforce/5544342190/, Public Domain, https://commons.wikimedia.org/w/index.php?

Humbolt earthquakes on November 5 and 7, 2011 that hit Kansas City and the surrounding suburbs.

These quakes occurred along the Humbolt fault line that runs from near Oklahoma City, up through Eastern Kansas and into Nebraska. This fault is near Kansas City and has caused destruction in the area in the past. In 1857 a large quake along this same fault brought chimneys down in Kansas City, but no one alive today in this area experienced an earthquake here before 2011. It is known among ufologists that so-called "Earthquake Lights" often appear just before or during an earthquake, and that these lights are often reported as UFOs. It is also a fact that more traditional UFO shapes and colors are reported before or during earthquakes. If these craft are piloted by beings of higher intelligence, these questions arise: Are the occupants causing the earthquakes? Could they be doing something to make an earthquake less destructive?

Perhaps UFO occupants are interested in the government sites in the area such as Lake City Army Ammunition Plant in Independence, the Richards-Gebaur Airbase in Grandview, the Kansas City Plant National Nuclear Security Administration faculty which produces nonnuclear material for the nuclear weapon arsenal,

Mystery Airship illustrated in the San Francisco Call, November 1896. Source: Wikimedia commons

high energy laser ignition systems, microwave microcircuit production and more. Or perhaps they are interested in Whiteman AFB, home to the 509th Bomb wing, home of the Stealth B-2 Spirit bomber, just 50 miles to the East of Kansas City.

A seemingly logical question would be: Why would UFOs would appear, brightly lit for everyone to see, if they were trying to hide? The answer may be that the sightings are accidental unintended, or whoever or whatever is piloting the craft do not care if anyone sees them, or are purposely showing themselves so people can see them. To what end we can only guess. Unfor-

UFO Reports in Missouri 2011-2015

	MUFON	NUFORC
Jan 1 – Dec 31, 2010	137	105
Jan 1 – Dec 31, 2011	253	140
Jan 1 – Dec 31, 2012	263	200
Jan 1 – Dec 31, 2013	171	110
Jan 1 – June 18, 2014	180	143
Jan 1 – Dec 31, 2015	159	100

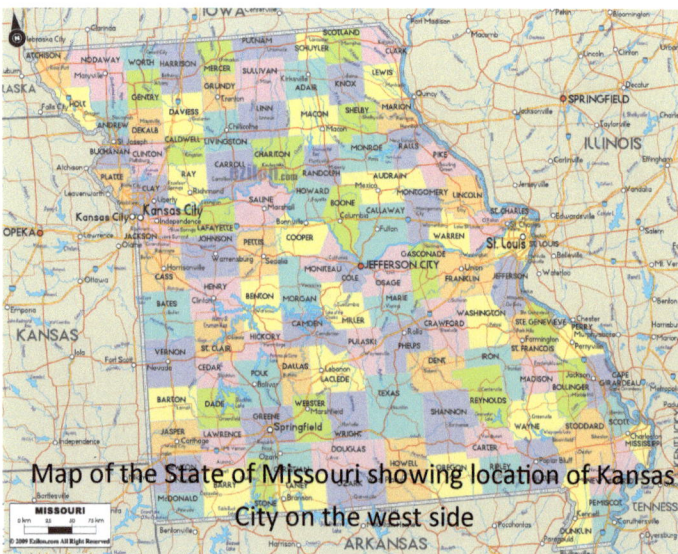

Map of the State of Missouri showing location of Kansas City on the west side

tunately, there are always more questions than answers when dealing with Unidentified Flying Objects or what are new being called Unidentified Arial Objects. What the UFO investigator's job includes is ruling out as many mundane explanations as possible - Sky lanterns, planes, helicopters, search lights, Venus, Flying Teams, nighttime parachutists, etc. until there remain no other explanation. Then we are left with a true Unidentified Object, which is what I present to you in this book.

A total of 47 reports remain unsolved and unidentified from the month of October, 2011. The unidentified UFO sighting reports include multiple reports of bright white, blue, orange, and dull black spheres at low altitude – even inches off of the ground- and large flying objects within feet of witnesses. Glowing spheres are the most perplexing of the reports, and occurred during the entire month of October and during other months. Of note are huge unidentified flying objects, two craft that turned on their sides and exited at a high rate of speed, and large diamond and triangular shaped craft that hovered at tree level up to 500 feet. And there are others. Some of the sighting reports have been identified as two airplane stunt flying groups (see further details following).

Debbie Ziegelmeyer, State Director for Missouri MUFON, Chief Investigator Joe Palermo and other investigators from Missouri, Kansas, and Arkansas assisted with the influx of reports by taking those which appeared to be identifiable. The bulk of the unidentified cases were handled by me because I live in the area. Several of the reports were submitted to me directly, and not through any association. Most of these were from pilots or law enforcement who so not wish to be identified.

What UFO investigators do is rule out as many explanations as possible, then we are left with a true Unidentified Fling Object, which is what I present to the reader in this book.

Retired Police Officer Reports UFO and Entity

The story begins in April, 2011 when reports of highly unusual encounters began to come in to my office. I present to the reader some of the most significant sightings in chronological order.

These reports came to me directly or through MUFON or the National UFO Reporting Center. Note, the table at the left does not include reports made to me directly.

April 5, 2011
3:15 am

John Doe (not his real name) was awakened at 3:15 a.m. in the morning by his female German Shepard, who insisted John follow her to the back of the house. The dog would not stop barking at something in the back of the house and wanted John to go to the back door. As he walked through the kitchen John saw a whitish/bluish bright light coming in the window through the door and the back windows of the house. John initially thought that the blinding light beam coming down into his back yard must be a search helicopter looking for someone. He soon realized that there was no sound from a rotor and thought that was odd. He became concerned. John is a retired police officer with years on the force so his instincts told him to go back to the bedroom to get his gun, which is always loaded.

He then returned to the kitchen and opened the back door, at that point the light just blinked out. John looked around the yard and looked up where the light had been a second before, and saw a very dark dull black triangle shape above floating approximately 40 feet above him. It had no lights on at that point, but blocked out the night sky. The object appeared to measure the width of his yard (50 ft.) and just floated, unmoving.

ILLUSTRATION: FOTOLIA.COM

John said "I heard a noise towards the back fence and looked in that direction. Then I remembered my flashlight and went to get it from the kitchen drawer, when I heard a sound in the hallway behind me. It sounded like an animal of some sort (two clicks), so I shined the flashlight in that direction for a few seconds but saw nothing. I then went back out to the back porch and shined the flashlight in the direction of a noise I heard outside and saw a very strange, short (approximately 4' tall), gray, humanoid looking creature with big eyes standing there looking at me for just a second or two. It just disappeared - it did not run. Then I looked up and the craft was also gone - I did not see it leave. I stood there, frozen in fear."

John's dog would not go outside, which is not normal behavior as she is very protective of John and his wife. John said that he is hard to scare after 30 years on the police force. He has FBI training as well and thought he'd seen everything, but this was not anything he had ever experienced.

"My first instinct was to call 911 and I went to the

phone, but after picking it up, I put it back down. I know everyone at the police department and even though we have received a few reports of UFOs in the past, I knew they would think I have lost my mind. But I haven't. I know what I saw."

John then locked the entire house up and went to the bedroom, not thinking about the noise he heard in the house earlier until the next morning. His wife was sound asleep, apparently sleeping through the entire event, and John did not wake her. He was too upset to tell her about what happened.

John thought he was gone only 10 -15 minutes from the time he was awakened by the dog until he returned to bed, but the time on his digital clock said 4:46 a.m. John can't explain what happened to the time he lost. The next morning John had a red sunburn on his face, neck, and arms as if he stayed in the sun too long, but there is no explanation for it. John told me that he just had to tell someone that this is real and that he is a 64-year old believer now after being a lifetime UFO skeptic. He told me he would call back, but never did. During the conversation I noted a distinct quiver in his voice, as if he was truly shaken up by the experience.

Interview in St. Louis Possible Cause for K.C. Sightings
May 13, 2011

I was in St. Louis to interview Ray Kosulandich about his encounters with extraterrestrials. Debbie Ziegelmeyer, State Director for Missouri MUFON, suggested that I listen to what Ray had to say. He had recently decided to go public with his story, which he had kept secret from friends and family for over 50 years.

We met at a racetrack early in the day in order to avoid crowds. There was no one around. I set up video equipment and began the interview but within five minutes, the battery, which was fully charged, drained down completely. We decided to move to another location 100' away. I again set up and interviewed Ray for about one hour. Ray told me about all of the different types of extraterrestrials he has been in contact with since the age of 12, and how he and his father would be abducted at the same time, or invited aboard craft together. The story is amazing and one that will be covered in more detail in a future book about Missouri incidents.

Towards the end of the interview, I suddenly had the feeling that there was a large craft above us and indicated that to Debbie, who had her camera in hand. I pointed to a specific area, then turned back to continue talking to Ray. Debbie pointed her camera to the sky and took several photos. Later, we found what looked like a clean hole punch in the clouds in the exact spot I suspected there was a craft.

Interestingly, right after this I saw a tiny 12" - 16" wide disc shaped craft hovering right over Ray's head. It was kind of funny—here he was talking about tiny beings, and there was a tiny craft right there that he was seemingly unaware of! Next, I saw three very small beings floating in the air to my left. They were just staring at me. Neither Ray nor Debbie saw them. *Note: I am able to see more of the light spectrum than most people so that may be the reason why this occurred.*

Now for the really strange part—I headed home to Kansas City but when I got to the Odessa exit I suddenly felt the need to grab my camera and take some pictures of the sky above me. So as I was driving, slowing down to about 55 mph, I stuck my hand outside the window and took a picture. Then I pulled over to the side of the road and looked up but could see nothing. I took another picture, and there is a UFO in the shot. It wasn't visible to the naked eye but the camera caught it. I had the feeling that whatever it was, it followed me home from St. Louis. So what happened next is all Ray's fault.

Three Witnesses See 125 Orange Orbs

May 14, 2011
9:36 p.m.

A Missouri insurance agent was travelling Southbound on I-435 in Kansas City, Missouri when five orange glowing objects in the Eastern sky caught his eye. The witness stopped to observe this phenomena, as the objects flew slow and jerky at about 20 degrees above the horizon. He pulled over to the shoulder to get a better view.

As the objects spread out and slowly began to disappear into the clouds above (low celling), a larger and brighter orange object moving at a faster speed appeared, about 1-2 degrees above the horizon, behind the tree line, ascended faster and split into five smaller orange objects, again jerky in their movements and without formation, ascended into the cloud cover at a an estimated 60 miles per hour.

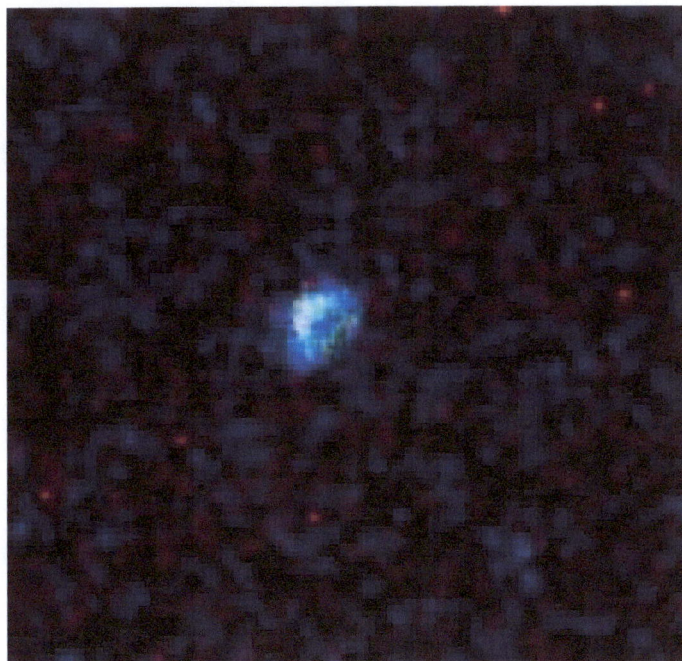

Witness photo of one orb close up (this one looks blue in the photo but appeared orange to the naked eye)

This same scenario repeated approximately 20 times over a 25 minute time span. The objects all came from the North (low on horizon) and headed South and then turned toward the East, and then up into cloud cover along the same path.

At one time, the witnesses observed over 15 objects in the sky. At approximately 9:58 PM when the largest ball going at a faster speed flew and split into 5 smaller objects and followed the same path as the previous objects the event seemed to stop.

The witness stated: "Once they all rose above the tree line, I would approximate their speed at approx. that of a helicopter. But the jerky motion didn't make sense, and the coloration from a side view made no sense, as the sun had been down or over an hour and the low altitude above the eastern horizon would pretty much rule out solar reflection. I would guess that over the period that I observed these objects, there were approximately 125 of them. I do not do drugs/ drink/or subject to illusions. I am observant, logical, and well educated. This happened, it was unexplainable, and a bit spooky. As this event lasted so long, with so many objects, I would presume that radar somewhere would have picked this up. If possible, I would like an explanation from some tower or authority trying to explain this to me. The initial speed of these objects at low altitude seemed to be about 100-120 MPH, but during ascent appeared to slow to around 60-80 MPH. The wind was about 15 knots, which might contribute to some of the unevenness of flight, but probably not."

Several other vehicles pulled on the shoulder of the highway to observe this event. The primary witness spoke with another man at the scene, who called a friend who lived at I-435 and 63rd street and asked him to go outside and see if he could see the objects. The friend did so and the next day the insurance agent obtained 13 cell phone photos from the third witness, which he forwarded to me.

Multiple orbs spotted with the naked eye—photo taken with a cell phone camera by a witness at 63rd Street and I-435. The orbs appear to be blue in the photo, but were actually an orange color. Note: the two bright lights are street lights.

The witness also contacted local media who covered the event in a television newscast. See the interview on YouTube at **https://www.youtube.com/ watch?v=39-YCkcXoxE**

I contacted the witness a few months later and asked if he remembered anything else about this event or if he had seen anything else unusual. He stated that he had been having nightmares and further sightings that he can't explain, and that it was affecting his work and personal life and that he wished he'd never seen anything. He would not discuss it further so I let it go.

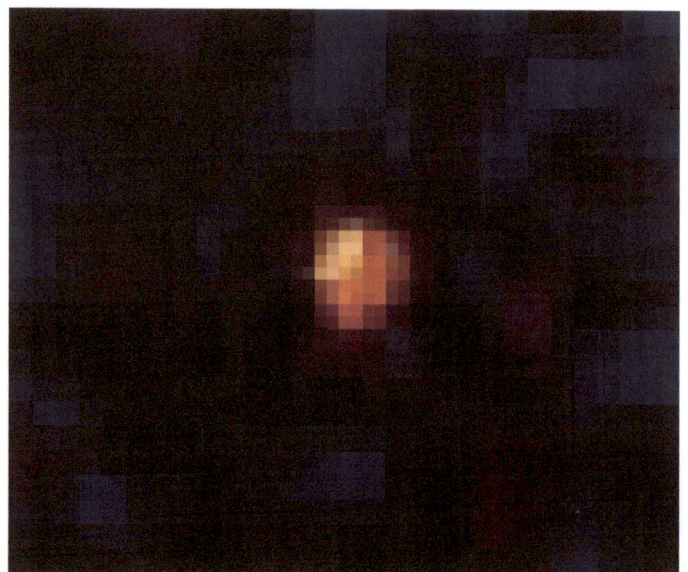

Close-up of one orange orb

Gigantic Saucer over Lee's Summit

Drawing of UFO over Lee's Summit on 5/15/2011 by Margie Kay and confirmed by the witnesses

May 15, 2011
4:00 am
Four witnesses

Felix Figueroa was driving home on Woods Chapel Road in Lee's Summit with his wife and three friends in the car after an evening out. At approximately 4:05 a.m. Felix saw a flying object he could not identify. He yelled at everyone in the car to look up in the sky. "Oh my God" was pretty much the reaction of everyone in the car. The object was very large and was not an aircraft any type the group had ever seen. As Felix drove closer in the direction of the craft it seemed to stay at the same height, yet either got further away or just got smaller and smaller.

After arriving at Felix's house three of the witnesses sat down to draw what they saw independently, and all came out with the same drawing of the ship.

Felix said "The strangest part of all this is that I live by an airport that single engine planes usually fly out of. For the last seven days I have seen multiple apache helicopters landing and flying around the airport. I have lived at my apartment for close to a year and have never seen anything other than single engine planes land there. We are all certain of what we saw and with the helicopters flying around all week I am convinced that there is an alien ship or something of that nature here close to where I live."

Several witnesses, including people attend Kansas City MUFON meetings and my secretary, who lives in Blue Springs, also witnessed the Apache helicopters. Apparently, they were coming and going from the Lee's Summit airport, which is a small airport. Residents said they've never seen Apache helicopters there before. When one landed, another one took off. The helicopters were loud and disturbed the sleep of many people. What were they looking for? Could it have been the large saucer-shaped UFO that appeared over Lee's Summit I the early morning hours of May 15th? And was this related to the orange orbs along I-435 Highway?

If a large UFO wanted to hide, it need look no further than Lake Jacomo, which is 45 feet deep at the center, and very close to the airport, or another lake nearby.

See the TV Coverage at **https://www.youtube.com/watch?v=39-YCkcXoxE**

On the same day of the strange disappearance of 10-month old Baby Lisa, who according to two child witnesses was taken by a "monster," the biggest UFO Flap in Kansas City history would begin.

Let me explain about the Baby Lisa Irwin case first. 10-month old Lisa Irwin disappeared from her home in Kansas City North on October 4, 2011. The case was heavily publicized and appeared on every news channel. Jeremy Irwin found the child missing at 4:00 a.m. after he returned from his late night job. Reporters spoke to many of the neighbors, including people who saw a man walking down the street with a baby, and two boys who say that they were up late and saw "a monster" taking the child away. The monster report is not on any of the news websites any longer, however, several KC MUFON members say they are certain that they saw it on a live news report. I mention it here only because of the strange timing of this event and the slim possibility that the case could be related to the UFO sightings that began to occur shortly after. As of today, the child has not yet been found.

Strange Gull Behavior at Longview Lake
7:30 pm

Witnesses reported seeing thousands of gulls fly away from Longview Lake near Blue Springs, Missouri between 7:30 and 8:30 PM on October 4. The gulls stayed in the air over woods and fields for over an hour, as if they were afraid to land on the water. The witnesses said that they live in the area and have observed the birds for years and this is something the gulls normally don't do.

A Close Encounter in Belton
June 19, 2011
1:24 am

Weather conditions: Storm with lightning

Names are changed in this report to protect the minors.

I interviewed Melissa S., age 33, by phone, then went to the site where I interviewed her and three teenagers who were witnesses on June 21 at 1:24 pm. I left at 2:00 pm and drove to the first sighting location to take photos. Present were Melissa S. age 33, D.E., age 14, K. R., age 13, and T. R., age 17. A fourth witness, T. E., age 12, was unable to be there but had drawn me a picture of what she saw and left it with Melissa.

At 1:24 am: N. and T. were looking out the patio doors watching the lightening in a rain storm when they saw a comet-like bright light shoot straight down from the sky over a house a block down the street near a four-way stop, which is south from their position. The object instantly turned into a disc– shaped object and hovered directly over the house for several minutes.

The girls screamed for everyone else in the house to come look at what they saw and continued to watch the object. They then saw three balls of bright white light shoot out of the craft and to the left (West) of the object. The other witnesses viewed the object for a total of approximately 10 minutes. Then the object faded slightly and zoomed straight upwards. This location is right next to the Belton Cemetery and there are two possible houses involved. .

The witnesses then moved to different windows in the house looking for the object in case it appeared somewhere else. After approximately 15 minutes, they saw it again, this time closer, and over a neighboring townhome approximately 40-50 feet away from their house. They watched this object for a couple of minutes, and everyone saw it clearly, only this time they were even more concerned.

The object shot four balls of bright white lights out in different directions, then faded out of sight. It did not move away, it just faded out.

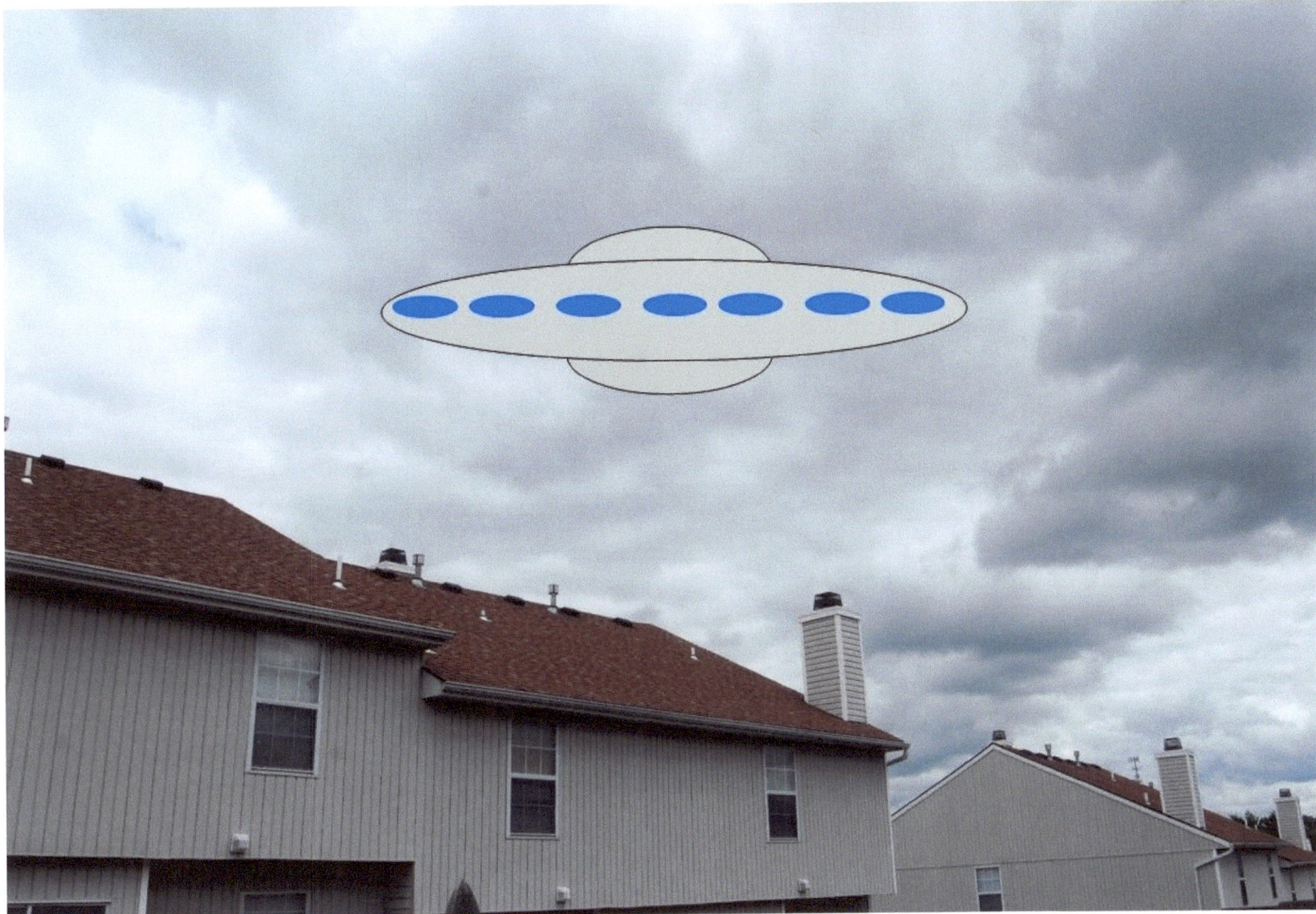

Above: Actual photograph taken at the location of the sighting with a drawing inserted based on the witness' drawings. I was standing in the back doorway of the townhome when I took the photo, so this object was extremely close to the witnesses.

Note: It is very rare to have multiple witnesses to events such as this.

Melissa then got on the internet to see where she should report something like this and found mufon.com. She considered calling the police, but didn't for fear of sounding crazy. She did not approach any neighbors because she did not want to seem crazy to them, either. Her husband was not home at the time, and does not believe any of the witnesses about this event. T. said that a friend saw the same thing from his house, but I have been unable to contact this other teenager to date.

Melissa said "I am a Christian and I do not believe in aliens. But I know what I saw and it was not a plane or helicopter. I am now questioning whether or not aliens exist." She was very agitated and concerned and kept asking me what it was that she saw.

I requested that each go child to a different part of the house and draw what they saw. Each produced a drawing that was very similar to the others.

Note: Melissa saw a white light with a blue tail shoot South to North in the sky while traveling on I-70 with her husband and 6-year-old two months ago. The light blinked out. She is now questioning whether or not this was also a UFO.

The witness also related a strange dream to me that she had a month ago. She and her son were driving down a dark road in her red car. She saw a bright light and a ship landed in a field with doors all along the side of it. They were open and had a stool in each door. A man was in one and a girl in another and they were both frozen. The girl had a scared look on her face. The other doors had no one in them. Then her car lost power and stopped and she saw a UFO overhead.

Melissa went to her garage in May to go to work, but her car would not start. A mechanic said that four different electronic parts are "fried." The mechanic said that he could not explain how this happened, but that some type of electrical surge may have caused it. The vehicle remains inoperable. This may or may not be related to the "dream" Melissa had about the landed craft.

What was going on here? One can only speculate. Was an alien craft hovering over someone's house to possibly abduct them, then realized that others saw their craft and moved towards them? If so, how did it know it was being watched? Is some form of telepathy at work here, or was this a military craft of some type with technology the public is not aware of?

Large Silver Disk

Downtown Kansas City
9-27-2011
12:15 pm

The following statement was taken by phone from a caller with an unlisted phone number. She stated that she just had to tell someone what she saw but did not want further contact. I felt that she was credible, so present this very strange incident to the reader even though there is little evidence. It is not unusual for government employees to report anonymously.

"I am a government employee working in downtown Kansas City, Missouri. When I went to lunch yesterday with two friends I noticed something hovering in the sky above us. It was almost transparent. It was a sliver craft with a whitish glow around it. There were no lights that I could see but it looked like it was lit up. I stopped and looked at it, and asked my friends if they saw something, but they did not. Then one person said 'yea, there is something there, but I can't see what it is.' I am not sure what that meant. My friends just wanted to go on to lunch, but I was shocked that I could see this silver disc-shaped UFO, but neither one of them

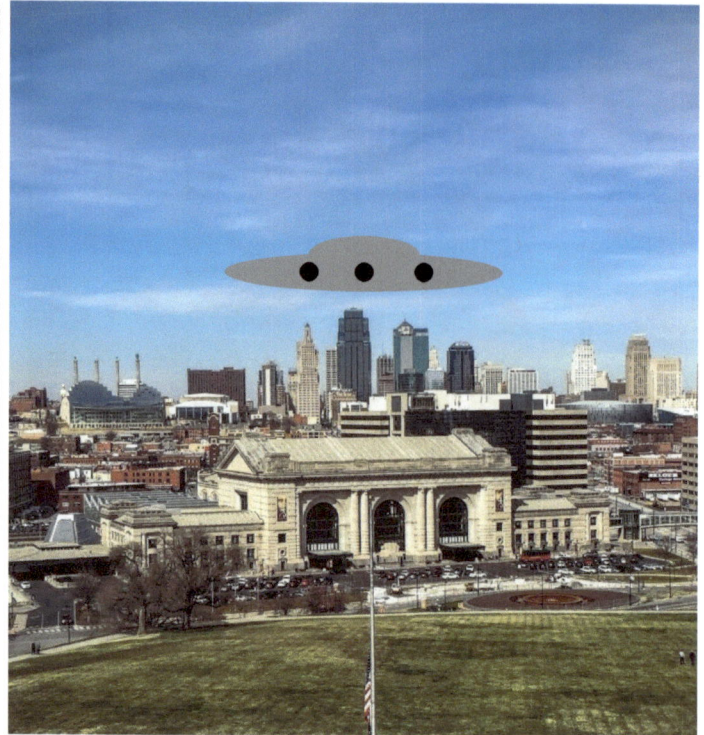

A portion of the downtown Kansas City skyline modified by Margie Kay to illustrate UFO

By Jordanbruening - Own work, CC BY-SA 4.0, https://commons.wikimedia.org/w/index.php?curid=48626090

could see it even though we were looking in the same location. I figured later on that maybe they just couldn't deal with the strangeness of the situation. I mean, a UFO in the middle of the day in a downtown area? I looked. around at other people, put they seemed oblivious to it. I've never seen anything like this and it really has me shaken. I can't give my personal information because I have a sensitive job."

This incident reminded me of a similar report from someone walking downtown over lunch hour, and my own five-second daytime sighting of a large craft hovering over downtown Kansas City which perplex me to this day. How can a gigantic UFO not be witnessed by hundreds or thousands of people? It makes me wonder if there is some technology that only allows certain people to see certain things, or if people who are intuitive are able to see things that others don't.

All Hell Breaks Loose

October 4, 2011

Highway 40 Encounter

Independence, MO

7:30 pm

This is one of the more detailed accounts by a witness to the events that occurred on October 4.

A woman named C.B. was driving home and as she approached her neighborhood she glanced down the street towards the clubhouse. Above it there was a big mass of lights hovering. The witness backed up and turned to dive toward the hovering object. She said "It just sat there! This gigantic mass of lights! As I got closer, I could see that it was a triangle shaped craft of some kind. It was only about 500 feet above the building, just hovering. It was black, but it looked like it was made out of mercury. I could see the whole underside of the craft as I drove slowly toward it. It had yellow-orange lights at each corner that pulsated and blue and white lights just inside the perimeter. They were very bright and they cast a bright light on everything below, but they didn't glare like our normal lights do. The light seemed to be a part of the surface."

Just as the witness arrived in front of the clubhouse, the object rotated and took off. It flew across the street and up over the trees, seemingly toward her house in a straight line. She goes on to say "I drove on down the street toward home and as I turned the first corner, there was a group of people running the same direction. I caught up with them just as I got to the last turn onto my street and one of the guys yelled, "Did you see it?" I said, "Yes!" He said, "It was a freakin' space ship!" I knew exactly what it was but it didn't really hit me until I heard him say the words."

The witness goes on to say "After I turned the corner, I could see it up ahead. It had stopped again just past my house. I drove down the street and pulled in the driveway honking my horn as loud as I could. I want-

Longview Lake

By Unknown employee(s) of the Army Corps of Engineers. - U.S Army Corps of Engineers website. http://www.nwk.usace.army.mil/Media/Images.aspx?mgqs=3Longview&page=1&stype=0, Public Domain, https://commons.wikimedia.org/w/index.php?curid=22052813

ed my daughter to come outside and see it so my family wouldn't think I was crazy or something. She said she didn't even hear the horn and I was really laying on it. Anyway, I jumped out of the car, forgot to turn the engine off and left the door hanging open because I wanted to go get her and my camera. As soon as I got to the gate, it spun around and took off again. This time it went almost straight up and it was gone in a fraction of a second. I have never seen anything move that fast in my life! Those other people had been running down the street right behind me, but after it disappeared, they were gone too. I don't know who they were or where they went but it was really weird. They were right behind me one second and gone the next! I looked up and then back toward them and they were just gone.

My heart was beating so fast by the time I got inside I could actually hear it. I think my daughter did too. She knew I wasn't kidding when I told her about it. I have never seen anything so strange in all my life. It was as big as three huge houses and it was just sitting there until I got close. Then, it went to my house and waited for me! That's how it seemed anyway. Later I looked all through my daughter's car for my phone and still couldn't find it. I went back in and then to my bedroom and there was my phone, plugged into the charger laying on my bed. I know I remember taking it off the charger that morning and putting it in my purse! I thought maybe it fell out and my daughter put it back on the charger but she swears she didn't. I don't know how any of it happened, but it did. Looking back, it really scares me. Did that thing know where I lived? And what was up with those people and my phone? I feel like I'm losing my mind but I am as sure about all of it as I could possibly be. I wish I could find those people but I'm not about to go door to door asking if the people who saw the UFO live there.

I filled out a report on the National UFO Report site but I didn't bother to call the police. What could they do? Probably just laugh. The next day my other daughter told some friends at work about it and two of them said they had seen something that sounded like the same thing within the past week. I wonder how many other people have seen them around here? Anyway, I spent several hours on the computer looking at new aviation technology and I can't find anything that's anything close to being like what I saw. Then a friend said something that made perfect sense. If it was an experimental government or military aircraft, why would they be flying it over a neighborhood full of people? Why would they hover for so long allowing us to see it? I know, without a doubt, it was not from this earth. I knew it the moment I got close enough to see it. I never denied the possibility of aliens visiting earth, but that just took all doubt away. It was unbelievably amazing! Now I'm reading that other's in this area saw something similar around the same time. Wow!"

UFO Over Raytown Water Tower 7:45 p.m.

One person called to tell me that he was driving on 63rd Street when he saw a woman stopped on the road video taping something. When he looked in the direction she was filming he saw a disc-shaped craft hovering just 15-20 feet over the Raytown water tower near 63rd Street and Raytown Road.

The object then rose up 15 feet, flipped on its side 90

The two men were mesmerized as they stared at the object and watched flickering red, green, and blue lights on the underside of the craft. The lights were each changing color, but oddly, did not appear to be individual colored lights, but rather one large light. When the lights changed color, they flickered at different rates. There was a blue aura around the object, and later a green aura.

As the men took photos and the cameras flashed, the craft moved towards them and emitted smoke, seeming to react to the flashes of light. The craft made some strange noises right after Lerone said "See this is not an aircraft." The object flew over their heads in a circle then flew down to the parking lot close to them just 30' away, and hovered a few feet off of the ground. Clarence felt a strong urge to move closer to the object and walked towards it, then got on top of a green utility box to get a better view and get pictures. Lerone admonished Clarence not do get any closer, but Clarence couldn't help himself. It was as if he were in a trance state, unafraid, and very curious.

After a few minutes the object made a very loud earsplitting thumping and low "wa wa" type sound. It then turned completely on its side with the top away from the witnesses, revealing the underside which was completely covered in lights. Lerone, from his position three stories above, remembers seeing 24 lights spaced perfectly from each other in a number of rows, and an appendage to the right of the craft with eight lights on it. Clarence's memory is slightly different, with lights in random patterns and the craft itself morphing slightly. Perhaps his vantage point from the ground provided a different perspective. The object continued to make a very loud terrifying clanging sound and a loud, low "waah waaah waaah" sound. Clarence described the sound as something that would be heard in a Science

Aerial View of Raytown

Photo: By Ichabod - Own work, CC BY-SA 3.0, https://commons.wikimedia.org/w/index.php?curid=30273817

degrees, and took off at "light speed" and out of sight. There was no noise, and there was a black contrail behind it. He said the size was massive. The witness said that some people came out of Wendy's restaurant to view the object, and that others were stopped on the road looking at the object and filming with video cameras but to date, no one has posted videos on the internet or filed a report with MUFON or me.

Huge UFO over Raytown Market 8:00 p.m.

The witness stated that "On Tuesday, October 4, 2011, at about 8:00 p.m., I was driving in the parking lot of Woodson Village Shopping Center at 63rd Street and Woodson Rd. in Raytown, Missouri, going west toward Apple Market, when I observed an unidentified flying object rapidly descending at a 90% angle from the western sky behind Apple Market. It appeared to move from the 90% angle into a horizontal position as if it were about to land on the roof of Apple Market.

At this time it appeared to be about 100 feet above the roof of Apple Market and about 100 feet in diameter.

The object itself was not actually visible to me, only the lights. There appeared to be a ring of approximately twenty regularly spaced bluish lights with a second smaller ring of what seemed like four flashing or rotating yellow-greenish lights as if they were perched some ten feet higher than the larger ring of lights. I think there were also three or four other yellow-greenish lights spaced between the two rings of lights, but it may have been the same three or four yellow-greenish lights appearing at different perspectives as the object moved at various angles and changed direction.

I almost discerned some black or dark structural aspects that gave the impression of a large disc weight as is used with barbells with a smaller disc weight centered on top of it. As the lights first came into my view moving down out of the sky directly toward me, I thought it must be fireworks, but it did not disintegrate and leave ash trails. Instead it got clearer.

I was really shocked and paralyzed with awe as it came toward me and did not seem as if it would stop. Then I thought it must be some kind of an aircraft about to crash, or perhaps a meteor, and you can imagine the thoughts that raced through my mind. I was quite calm and thinking logically through all this. I thought to grab my phone to call and say a final goodbye to family, but did not feel I would have time to make such a call. That's how fast it was moving and how close it came. I was not afraid, but nonetheless relieved when it leveled off and hovered over Apple Market instead of continuing on towards me. I reached for my camera, which I usually keep in my car, and turned it on, but the object suddenly turned due north of Apple Market and flew over the subdivision immediately north of 63rd Street.

It appeared to fly just barely above the treetops and disappeared. In a second or two it reappeared partially obscured from view by the trees. The entire sighting lasted about four minutes, long enough for me to get a really good look at it. I just sat in the car in the parking lot for a few moments trying to figure out what had just happened. The lights were so beautiful, like a colorful constellation, a carnival ride, fireworks, or Christmas tree lights. Once they did not crash into me, I really hoped I could see them again.

As I mentioned before, I did not feel frightened, only exhilarated. The object had not made any noise as far as I could tell. I have seen airplanes and stealth bombers and helicopters and it did not seem to be any of those, unless one were painted black or very dark and had a ring of lights attached to the bottom of it. I thought for a moment it might have been a police helicopter searching for a suspect in the dark, but it was way too big and did not have search lights, only the lights described earlier. Plus, those tend to be very noisy, and this object was not. I considered the possibility that it might have been some kind of projected holographic image or special effects for a sci-fi movie being filmed. If it had been a pre-Halloween prank, it was a good one!

I looked around to see if anyone else had seen the event. There were only a few cars in the parking lot. I only saw two people and they were outside of Apple Market, one was just coming out of the store apparently oblivious to what had just occurred. The people inside the store I saw through the window seemed equally oblivious. There was a man walking from his car toward the entrance of Apple Market with his head bent downward and he had been parked so close to the building that he likely would not have seen the event. The blinds at the window of Fantastic Sam's were drawn and I did not see anything unusual through the glass door. The other stores were all closed. Only the bar was open, and I was not about to go in there and announce what I had just 8seen. I sat in the parking lot until 8:08 p.m., and then I began to drive in north on Woodson in an attempt to get a glimpse of the object again. I drove throughout the subdivision north of Apple Market but did not view the object again. I called only a few close family and friends. I was really concerned that if I told anyone else they might think I had become mentally unstable or was hallucinating due to some physical illness or fatigue.

I am convinced that what I saw was very real and not some hallucination or figment of my imagination. I had considered that it might be some kind of spiritual phenomena, and as appealing as that thought may be, I don't think it was. I am highly doubtful that the object was made by aliens from outer space, but I would not rule out that possibility. My best guess is that what I saw was some secret military project of the U.S. or some other country, say, China. I heard on the radio today that numerous independent reports of sightings of unidentified flying objects in the Kansas City, Lee's Summit, and Raytown areas had been made. That feels like a validation, although I don't need one to satisfy myself."

Large Craft over Major Highways 8:10 pm

Several witnesses reported seeing a very large craft hovering over 291 Hwy and I-70 Hwy. In all of these cases, I first thought the object witnesses reported was the EAA Flying team, however, I was unable to attribute the flying team to several reports. There had been an accident on the highway and it was backed up for miles, so many people were stopped and took notice of a large object in the skies.

In all of the unexplained reports, the witnesses were with at least one other person who corroborated their story, they all were insistent that it was one large craft, and that the craft hovered for several minutes over the highways. So the flying team had to be ruled out since their planes are not able to remain stationary.

In one instance, a couple got out of their car as did others who were parked on the road. The couple said that over 100 people were standing outside of their cars looking at the unidentified flying object in the sky above them, and that some people panicked and got back in their cars.

One witness describes his experience:

" This is Nuts, I saw a UFO when I was 14 with my mom a long time ago in the distance, but tonight was nothing like that. This was the craziest thing I have ever seen, I was so close to it, I can't believe it.

I was driving down interstate I-70 westbound and came up to the 291 Hwy intersection that was backed up due to an accident on 291 Hwy south. I looked up and saw this hug gigantic object beaming with white bright lights all over it, and it was hovering in mid-air right above the intersection.

It was hovering about 500 feet above the highway. I pulled the car over and watched it just hovering. I rolled the windows down to hear it but there was no noise whatsoever coming from the object. Total silence—just eh cars passing by. I was freaking out and yelling. I could not believe my eyes—or what I was seeing. My family was with me and we were all freaking out. Within about two to three minutes of hovering it turned south and started at a straight path going south. I tried to take pics but my camera on my phone couldn't pick it up enough. I saw the rear of the craft and it was huge. The back had three separate box=shaped glowing lights that looked as though they were spreading apart.

It headed along 291 Hwy south, but was more flying over the Cliffs Homes area direction. 291HWY south was closed off by police due to the wreck, so I couldn't follow it but I did head down 40 Hwy west off of 291 Hwy and then went south on Lee's Summit Road, but did not see it due to too much traffic and trees. The

shape was like a small pyramid, dome on top but covered in bright white lights, lots of them, and blinking blue lights near the bottom. But the bottom view was a total V shape, no doubt about it. Then as I saw it from the rear I could only see three bright white square shapes in a row, but not straight in a row—all three rear aqua lights were in different angles. This has changed my life going forward.

I have never seen anything like this nor have been as close to one. I mean, this thing was hovering right outside the car window above me at a 45-degee angle in front of me. Just pure crazy… amazing! I never thought I would see one again in my life. "

Note: The fact that over 100 people saw this object and only a few reported it shows that most people don't report seeing UFOs. If everyone who saw a UFO did report it we'd probably get about 95% more reports that we do now.

Large Craft over Downtown Lee's Summit
8:20 pm

The witness called me directly and said "I was driving towards Downtown Lee's Summit after work last night when I saw a very large and low flying craft with three red lights on each side and several blue lights on it moving very slowly in a North direction. It was the size of a large cargo plane. I could not see any wings or tail and there was absolutely no sound. I called my buddy and he saw the same thing from a different location (his report is filed). I will submit a drawing. I did not have a camera with me at the time." Note: The witness was at work while I spoke to him and he didn't have much time to talk. We did not connect again. This is likely the same object that others reported at 8:10 pm.

Gigantic UFO and Possible Abduction
8:30 pm

This is the biggest case out of all I investigated on the night of October 4, 2011, and is one of the biggest cases I know of in recent history.

Clarence Williams, Lerone Pryor, and Lerone's wife Buddy were at an apartment complex near 63rd Street and I-435 Highway in Kansas City, Missouri when several children playing in the parking lot pointed out an object in the sky. The children, aged 10 to 14, were a few yards from Clarence's position.

Clarence pulled into the parking lot to park before heading to meet his friend Lerone, but he stopped when he saw an object in the sky, which was pointed out to him by the children. Lerone and Buddy came out on the third floor balcony after Clearance yelled for them. The three adults watched a UFO moving in a circle over a field near the complex northwest of their position.

They watched as the UFO moved to the South, changing shape several times as it seemed to scan a vacant lot and mining area with a light. A green colored beam emitted from the object, but the light from the beam stopped in mid-air and was not visible on the ground. The object ejected several green orbs of light. The children ran inside their apartments because they were afraid of the object.

Lerone Prior and Clarence Williams indicating where the bottom of the craft was located as it hovered off the ground

Above: Photo of object as it approached the witnesses.
Photo: Clarence Williams with cell phone camera

Left: Photo of object as it approached the parking lot (tree and green mist in foreground)

Photo: Clarence Williams

fiction movie.

As Clarence stood on the green utility service box approximately 20-25 feet from the object, Lerone, who was standing on the third floor balcony of the apartment approximately 30 feet above the ground, yelled to Clarence as he watched a green mist come out of the object and down over Clarence. The mist seemed to put Clarence in a trance-state and he did not hear or respond to questions. At one point, Lerone yelled several times to Clarence "Do you see that beam of light?" but Clarence did not answer or look in Lerone's direction. Clarence does not remember this incident, but does remember seeing a set of stairs come down from the bottom of the object. He took photos, and one photo does show what looks like a staircase and amazingly, two entities standing next to the stairs.

Buddy saw Clarence fall down during their sighting but Clarence does not remember that, either. Clarence said that he remembered feeling at peace and wanting to go towards the craft, but has no memory of the trance state or of anyone yelling at him. Both men obtained photos, with one photo taken from the balcony aiming down at the top of the UFO as it hovered below.

Lerone used a Canon digital camera and Clarence used his cell phone to take pictures. Oddly, there is a 20-minute gap between photos on both devices. Both men claim to have been taking photos constantly during the entire sighting event and cannot explain this gap. I examined the photos and the date and time stamps and there is indeed a gap in time.

The witnesses then watched the craft move up slowly to approximately 500 feet from the ground, then suddenly fly very quickly to the north of their position. As it did so, the lights on the craft went out one by one, making the craft no longer visible. The path it followed was north along I-435 Highway.

After the craft left the area, Clarence suddenly had the compulsion to go to Loose Park in Kansas City to watch for the object to return, against the advice of

Lerone. Clarence woke up the next morning at The Point, which is an area overlooking the downtown Kansas City Airport at the Missouri river. Clarence does not remember how he got to that location or why he apparently slept there.

The craft was "huge," according to the witnesses. Oddly, Clarence and Lerone described the object somewhat differently. For instance, Lerone and his wife saw a staircase come down from the craft and watched it for a few minutes but Clarence only saw it for a couple of seconds. Clarence thought the object had a fluid movement, as it the craft were alive, while Lerone thought it was a solid object.

There are missing photos and video from the camera and cell phone. Lerone put his flash card in his television so I could see the photos I took some screen shots with my HD Bloggie camera just so I'd be sure to have them. I then took the flash card from Lerone's camera and drove to my office immediately and put the card in my computer. The trip to my office took only 20 minutes, but when I opened the file all photos of the object, which I had just viewed were missing! Luckily, the photos once again return after I gave the card back to Lerone.

Implants?

Shortly after the event, a small hard object appeared under Clarence's skin on his neck, and remains there today. UFO investigator Larry Cekander first scanned the object with an Electro-Magnetic Field meter at the Kansas City UFO Conference in 2012, and found that the object does emit EMFs, which is unexplainable. Later, Clarence found more possible implants in different areas of his body after seeing strange lights and I scanned the objects with an EMF meter a number of times and they always emitted an EMF field.

Second Site Investigation:

In April of 2012, I returned to the site with several other investigators including MUFON State Director Debbie Ziegelmeyer, Investigator L. J., and my friend and police officer Corey Pearce. I take Corey with me on unusual cases since he has police training on crime scene investigations. Debbie and I met for lunch at a restaurant nearby while waiting for the others to arrive, and noted a couple who entered the restaurant who sat in the booth right behind me, even though most of the other booths were empty. The man sat with his head turned to the right as if he were listening to our conversation the entire time. As soon as we left, the couple left immediately, before finishing their meal, and

Right: Close-up of center large light on the object.

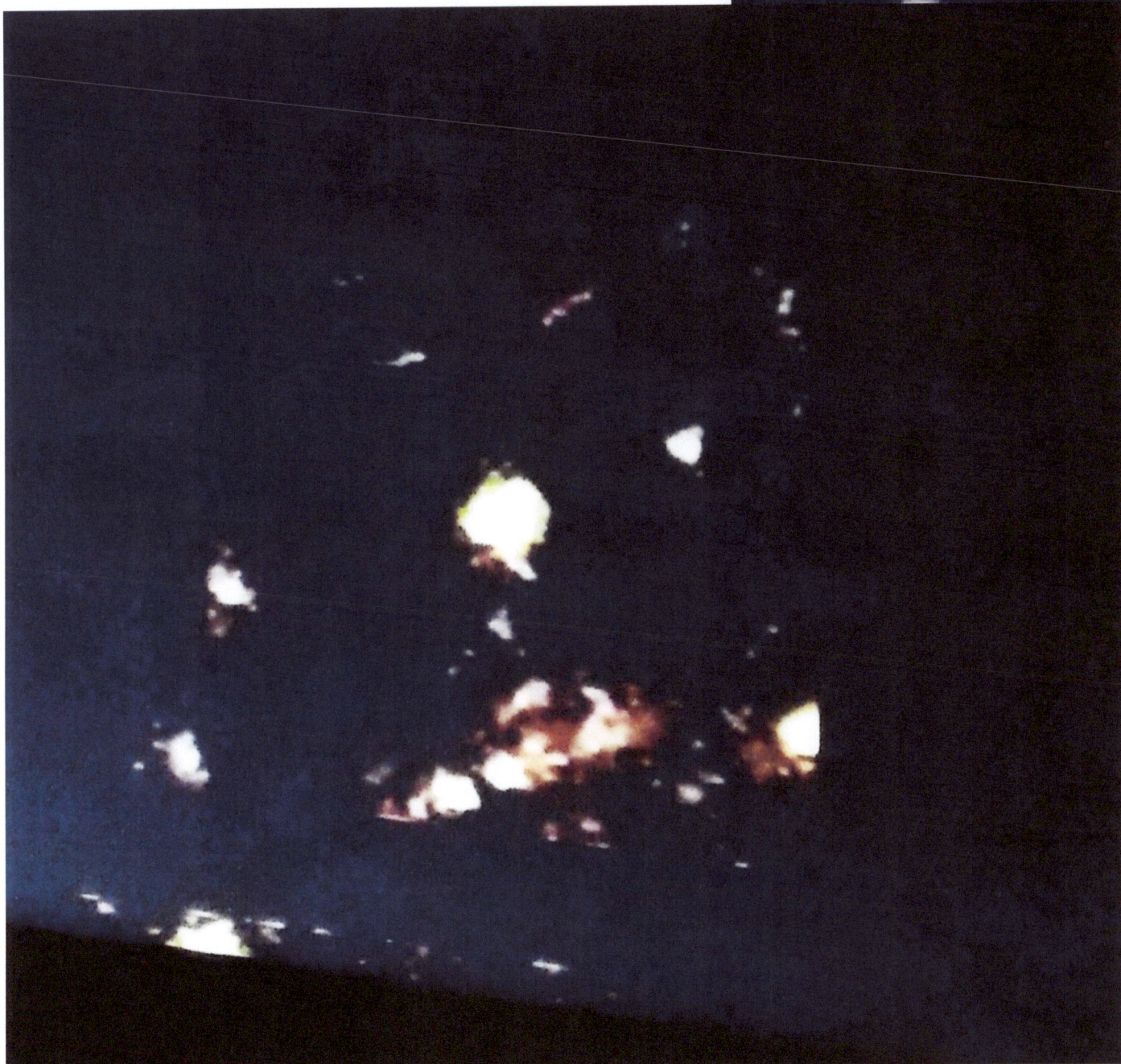

TV screen shot of UFO very close to the witnesses taken from the balcony. Photo: Copyright Lerone Pryor

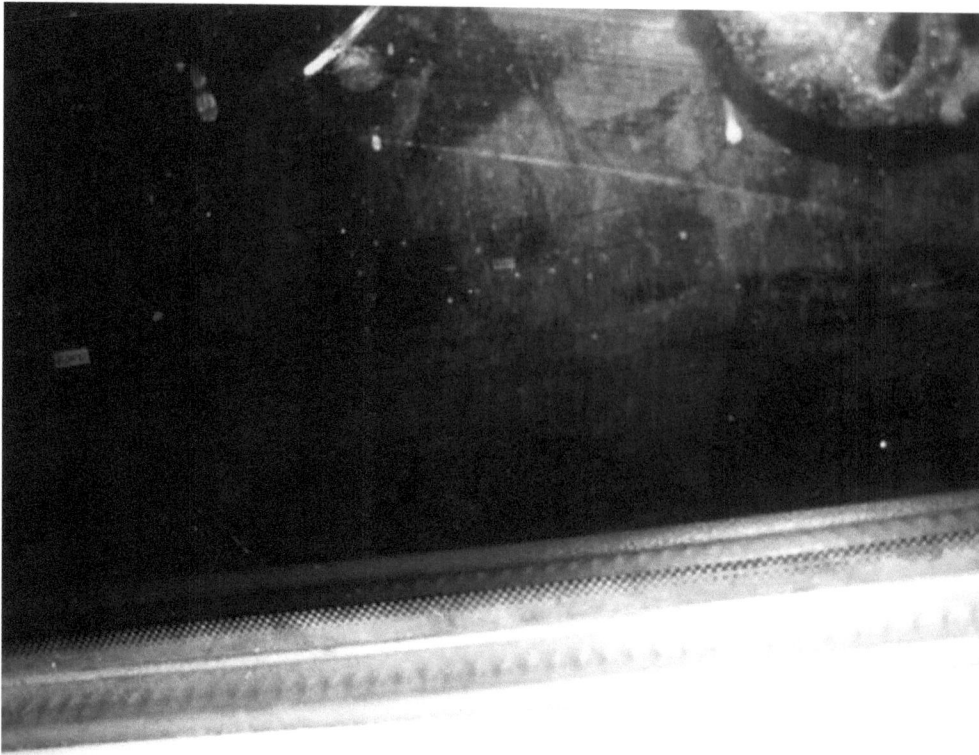

Aliens at a Gas Station?

About a year after the sighting, Clarence and his daughter stopped at a gas station to fill the car with gas when his daughter saw something standing to the right front of the vehicle near the windshield. She snapped this photo with her camera (left). The two believe that there is an alien head tilted sideways looking inside the vehicle.

Photo : copyright Clarence Williams

followed us out. Debbie and I stood around the corner in order to see if they might have been following us, and sure enough, the man came out first and was looking frantically around the parking lot as if trying to find someone. He cursed out loud. We are not sure exactly what that couple was doing, but it was very strange.

We then met Clarence and Lerone, who took us to the area where they thought the UFO had been doing a search with the green beam of light. Debbie, Larry, and Corey walked down a steep wooded hill to investigate an area that looked like a crop circle formation. Clarence and Lerone found the area after looking for evidence in the location were they first saw the UFO. The investigators did find three depressed and swirled circular formations in the tall grass, however, since it had been several months since the incident it was useless to obtain samples of the grass. I called crop circle expert Nancy Talbott of the BLT Research Center and asked her what needed to be done for a crop circle investigation, and we did as she advised, however, it had been too long since the incident to take samples. Photos and measurements were taken. The in-

vestigators also found a different location where something burned the ground, but there was no way to tell what caused it. We wonder what and who could have caused the circular swirling and depression of the grasses and why this anomaly was in the exact area where Clarence and Lerone and the other witnesses saw the flying object go to and possibly land.

Government Interest?

While the three investigators were searching the lower area, I remained at the top of the hill standing outside my SUV parked in a cul-de-sac, while communicating with the team via cell phone. After approximately 40 minutes, I returned to my vehicle and shortly after a large black new model dual-cab pickup truck with dark tinted windows pulled in close to my vehicle, blocking my exit in any direction, and stopped. I could see a large black man dressed in a suit in the driver's seat. He just stared straight ahead and did not look in my direction. He was talking to someone, possibly via an Bluetooth ear piece. The man drove away after a couple of minutes, speeding down the street and out of sight in seconds before I could get the license plate number. I felt that it was a definite message. I called the team back and Corey ran back quickly, but too late to see the vehicle.

We then returned to the parking lot and apartment complex to obtain photos and further evidence. We used a UV light on surrounding vegetation and an EMF meter on objects in the area. Oddly, I noted EMF readings on the wood fencing on all four sides around the garbage area, which was very near the location of the UFO when it was near Clarence. There should have been no EMF readings in that area since there is no electricity, power lines, or electronic devices that would cause a reading.

We interviewed a 14-year old girl during this visit and she did remember the incident clearly. The girl said she saw an object, pointed it out to Clarence and Lerone, and watched it move over the mining area. She became frightened and left the scene shortly after seeing the UFO to go inside with her family. She did not look out the windows or go back outside.

The Strangeness Continues

After the event and investigations, Clarence and his wife began to see strange orbs and lights inside their home. They were able to capture photos of some of these objects, which seemed to operate under intelligent control. They were all self-luminous. No explanation could be found although the couple looked everywhere for a source of the lights. The orbs appeared on numerous occasions.

Clarence was thinking about the strange events one day when he heard the name Fisidio. He believes that the name has something to do with an entity who is responsible for these strange occurrences.

It should be noted here that Lerone is now divorced. His wife could no longer take the strange events that seemed to plague him.

As of October of 2016, Clarence and Lerone continued to see strange objects in the skies day and night.

During a nighttime visit to Swope Park in the spring of 2016 Clarence, Lerone, and Lerone's new girlfriend all saw an object with bright white lights on it hovering and moving about a hillside nearby. Soon several helicopters arrived and circled the same area as if looking for something. The object, however, lowered into the trees and turned off its lights just prior to the arrival of the helicopters. The helicopters left, seeming to give up the search, and shortly **after** the object reappeared, seeming to scan the area for something.

The three watched this for a couple of hours, then left for home. The next morning Lerone passed the area on his way to work and saw that the hillside where the object had been was on fire. He heard fire trucks headed for the area, but he did not investigate further since he had to get to work.

The men talk about an X pattern that will appear in the skies if Clarence requests to see it. Clarence started seeing X patterns in the skies shortly after the close encounter. These patterns appear to be contrails or clouds in a distinct X formation. However, if Clarence sends a telepathic message to Fisdio that he wants to see an X pattern, one will appear for him and whoever is with him to see.

Healing Abilities

Shortly after the event Clarence got the idea that he could heal. Lerone's child became very sick with an asthma attack and had to go to the hospital. The child was sent home with little hope of recovery, so Clarence worked on him, sending healing energy. The child woke up the next day, completely healed with no trace of asthma, and the doctor cannot explain it. He has healed other people, too.

While a co-worker was struggling with a vehicle that would not turn over one day, Clarence suddenly had the thought that he could fix the car by touching it. He put his hands on the vehicle for a few minutes, then asked the co-worker to turn the key, and when he did so the car started immediately. After that incident Clarence has been able to repair several vehicles using the same technique.

Clarence's life has changed dramatically since the incident. Before the close encounter occurred, he was angry and depressed. He was often in arguments and fights. According to his wife, Clarence has changed completely and is now a calm, caring person. In fact, just before this incident Clarence was shot by a man, and he fully intended to retaliate, but after the sighting Clarence went to the man and forgave him, saying that he wanted to stop the violence.

Clarence now feels that he has a new purpose in life, which includes stopping violence, and that the gift of healing was provided by the UFO craft occupants.

Questions that remain unanswered:

Why is there missing time on two devices?

Why were there EMF readings on <u>wood</u>, five months later?

What are the objects in Clarence's Neck and other areas?

Why does Clarence now have healing abilities?

Were we being watched by some government agency?

Note: At least fifteen witnesses describe a similar disc – shaped craft on the same night. Some describe craft with multiple lights and appendages coming off of it, and in some cases, other smaller craft docking and un-docking with it. One police officer in eastern Jackson County came forward with a 15-minute video of a UFO taken from his patrol car dash cam which is now mysteriously missing from the police video storage. This officer now wants to be a MUFON Field Investigator. A professional photographer, who also witnessed a UFO on that night has volunteered to analyze witness photos.

A more detailed account of this incident and both of the witnesses is forthcoming in a new book by UnXMedia.

Radar Reports

A radar report for the night of 10/04/2017 with several photos was obtained through MUFON. These reports show that one target moved into three levels of radar in a downward direction during a 6-minute sweep.

The radar report also indicated that a **single target remained at 8,500 feet and was stationary,** which is extremely odd.!

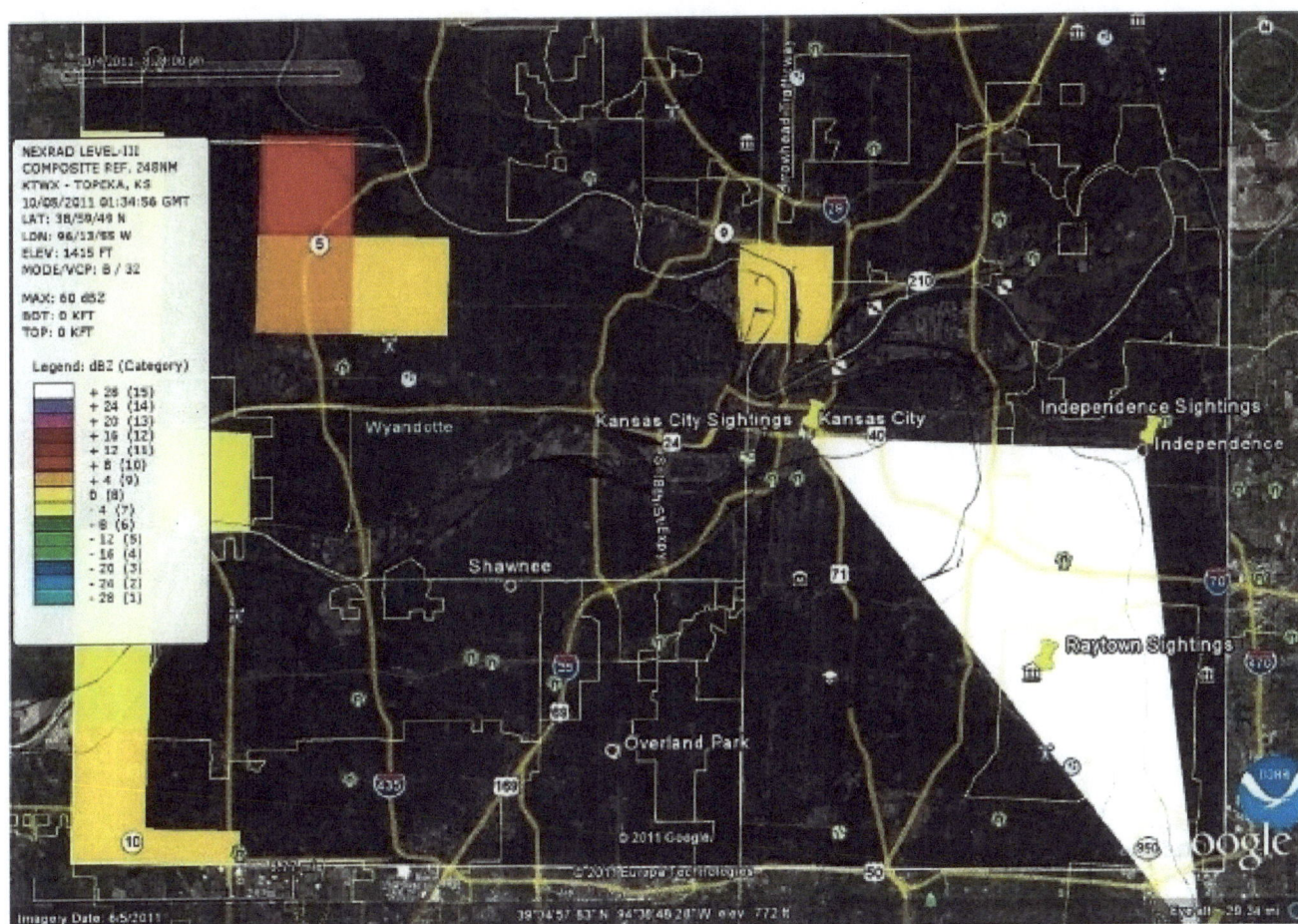

Missouri Sightings 10-4-2011   October 9, 2011

(V)   Here is a Composite Reflectivity image from 8:34pm on 10-4—2011. The triple-block image (red, orange, yellow) indicates that the target moved into 3 levels of the radar beam during the 6 minute sweep. Target may be coming down from a higher altitude, thus easier to see by the distant radar, (red), or a target going down and around losing echo strength on the decent. The single-block image just north of Kansas City, does not seem to move at all remaining at approximately 8500 feet , 9 miles from the sighting area. In a cosmopolitan area such as this many types of radar hits are going to be possible thus making the investigators job more exacting.

###

A second image from the radar report at 8:20 pm indicates that there was an object in the sky approaching the Kansas City area from the east.

Missouri Sightings 10-4-2011

October 9, 2011

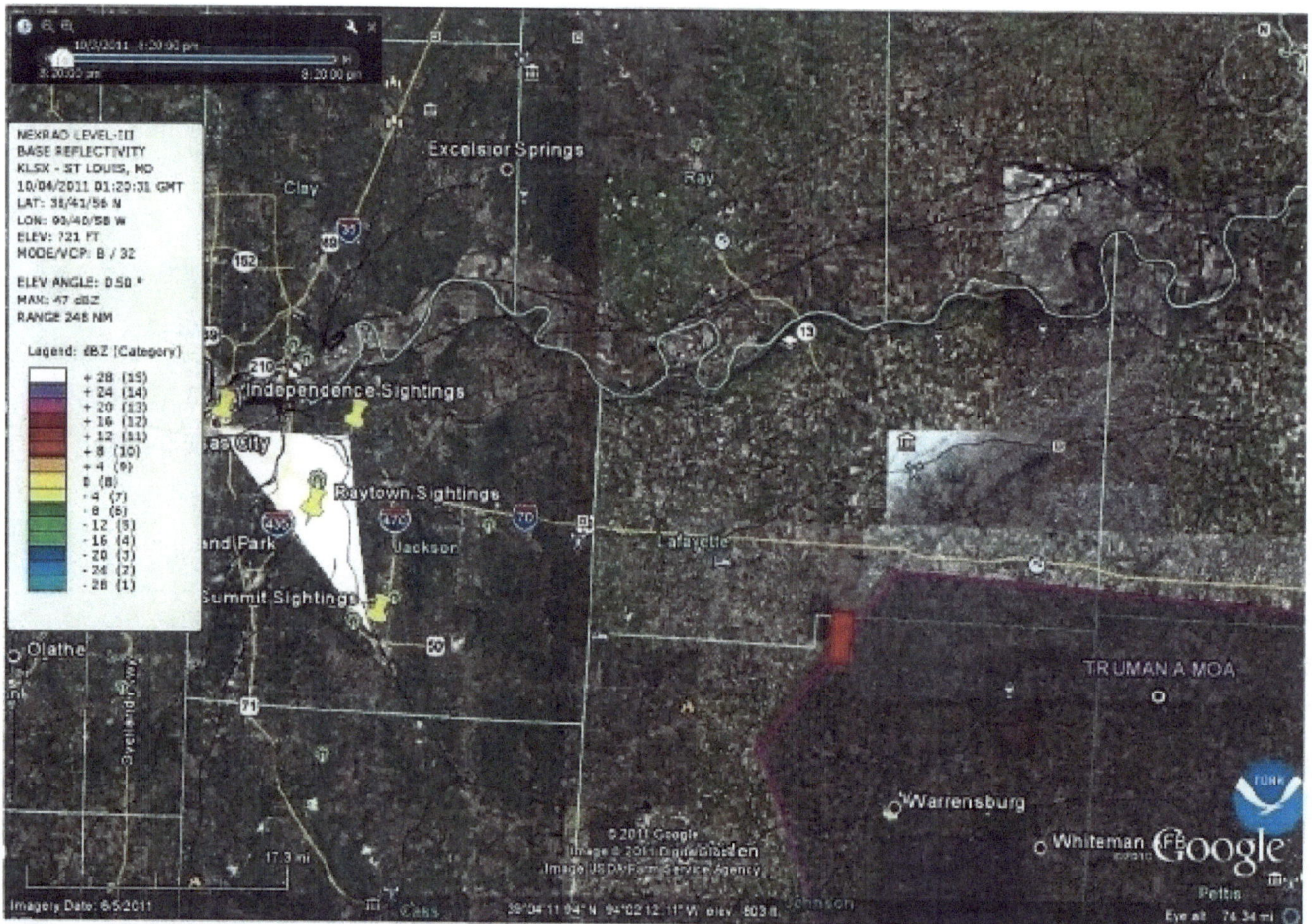

(Q) Target is 35 miles from the sighting area over the Truman MOA. This target presented at 8:20 pm on the 3rd of Oct. Approximate 38k feet in altitude. This was a good strong echo. Base Reflectivity at elevation angle of 0.50°.

Note: the above says the 3rd of Oct, but was actually the 4th.

10:50 pm
Gigantic Diamond Shaped Craft over I-70

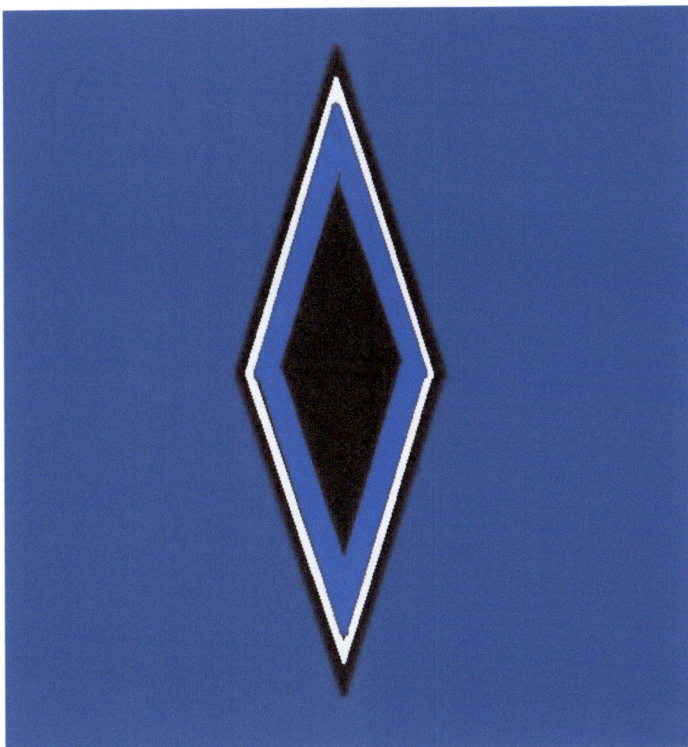

Drawing of diamond-shaped craft over I-70. Drawing by Margie Kay as described by witnesses.

This event was reported directly to me via phone. The witness said "I was driving north on 435 Highway from Overland Park to Kansas City North on my way home from doing a video shoot for a commercial. I am a professional videographer and so is my wife, who was with me. As we came around the curve at the 470 interchange I noticed a large diamond shaped craft moving very slowly along my right side (east) approximately 1000 yards from the highway and approximately 500 feet in altitude. It was 9:50 P.M. at that time. The object had a continuous white light all over the exterior sides of the bottom of the craft, and another continuous blue light just inside that one. It was a beautiful medium blue color that I cant describe, and while the white and blue lights were very bright, they had a soft glow to them, too.

The dark craft (I think Grey) reflected the lights. It appeared that the top of the craft was kind of domed in the same diamond shape, and the bottom was flat except for the lights. At first I thought it was some kind of blimp but soon realized it wasn't. I've never seen anything like it in 33 years! It continued along the same path for a few minutes until as I approached the I-70 exit the object stopped and hovered for approximately 20 seconds. I was driving and told my wife to watch the object. She was getting very excited about it and watched it the entire time. I had to keep watching the road and slowed down to 50 mph so we could see it better, but there was no place to turn off.

There were no clouds out and it was difficult to tell what the altitude was, but it was flying very low, too low in fact. A Commercial airliner flew over this craft headed SE to NW, but was flying at high altitude. Then, just as we passed the I-70 exit, the object turned to the west and headed that direction, again very slowly, and passed over the highway just as we went underneath it. My wife was saying oh my God, oh my God over and over again. We saw two cars in front of us slow down and it looked like they were watching it, too, but again, there was no safe place to pull over and at this point, I did not want to stop anyway for fear of what might happen.

The car radio, which was OFF, made a crackling noise for a few seconds, then a burst of some type. My wife said "what the hell was that? She was trying frantically to get her cell phone camera to work when the phone went completely dead. She had the phone charged up fully before we left the house earlier before the video shoot, and only used it a couple of times so that battery should not have died! My camera was in the bag in the trunk and I couldn't get it. Then, just after the craft crossed the highway, maybe 100 yards past the highway, it took off at a very high rate of speed directly west towards Kansas. I'd guess that it was traveling at 750 MPH or more, at an instant without any time to accelerate like a normal airplane would.

There were no features or lights on this craft that would indicate it was a common aircraft. It disappeared from view within five seconds! When we got home we called some friends to tell them what we saw, and they told us that they saw some strange white balls of light around their tree line at their house earlier in the evening and their neighbors saw them, too. A couple of days later I heard about all of the UFO sightings in the area on the news and decided to find out where to file a report. That is when I found Margie Kay's website and called her.

I am a pilot and my father is a pilot. He has his own Cessna Caravan 208B that I've flown with him in for years as a co-pilot from the time I was 14 years old. I can tell you that this object was no plane. If it was one of ours it was flying illegally without the proper warning lights and it was below 1,000 feet. I'd really like to know what it was!" Due to my father's work and my work, I have to remain anonymous."

I called MCI (KCI) airport and inquired about any anomalous radar readings on the evening of the event, and they said they knew of none. I then filed an FOA request but received no response. The witness sounded credible.

I also received two other reports from witnesses who said they saw a large craft further north along 435 Hwy at approximately 11:00 pm. These were near 152 Hwy in Liberty. These three sightings could have been of the same object that appeared at I-436 and 63rd Street.

The Investigations

I began working on these cases which occurred on October 4, 2011, and enlisted the help of other investigators due to the volume of reports. We found the following:

The B-2 Mission:
Stan Seba, a MUFON Star Team Investigator in Kansas, called Whiteman Air Force Base near Warrensburg, Missouri and reported the following: "After a long talk with the Public Information Duty Sergeant at Whiteman AFB, I found that the B-2 bombers of the 509th bomb squadron have been training for the past five nights and are on yellow alert. They will continue training for at least a few more days/nights.

The B-2s took off from Whiteman just after dusk and their training missions took them right over Kansas City, Missouri. Due to the fact that they were flying low over a populated area, the B-2s flew with their navigational (landing) lights on. The pictures some of the witnesses provided match those of B-2 aircraft. All B-2 aircraft reported their positions as over KCMO at approx. 20:00 to 20:15 hrs., or between 8:00 and 8:15 p.m. On October 4, 2011.

As for the lack of noise, if the witnesses were watching the aircraft approaching they wouldn't have heard anything. Unless they were almost directly behind the aircraft they would have heard very little noise. Also, the Air Force did have "VIP" observers to these exercises over KCMO in helicopters."

Radar Report:
A radar report obtained by MUFON for October 4 indicated that there were several targets in the area at the time of the sightings which could be attributed to planes or other objects. Oddly, and of note, was a stationary target at 38,000 feet nine miles north of the sightings for approximately 20 minutes beginning at 8:00 p.m. and ending at 8:20 p.m. This remains an unexplained mystery. Could this have been the large craft that was seen by multiple witnesses, or was it a "mother ship?"

The Flying Teams:
Debbie Ziegelmeyer contacted the *EAA Flying Team* and found that the six aircraft flew in wing formation over Lee's Summit, the Chief's Stadium, then over I-70 and back to the Lee's Summit airport, where they landed in pairs. The flight was between approximately 7:00 P.M and 8:00 P.M. This was a demonstration or audition for the Kansas City Chiefs for a night time fly-over planned October 31. The Flying Team uses multiple strobe lights on their aircraft and flies in close formation, which could be misidentified as one large craft. It certainly is a strange sight, and one that would be notable.

Airport Reports
I contacted the Lee's Summit Airport, and confirmed the flight time for the EAA Flying Team. The contact at the airport stated that they had received a number of odd reports of UFOs the past two weeks prior to the October 4 event. One report involved a low-flying helicopter at 1:30 a.m. over houses when no helicopters should have been flying, and another report involved low-flying balls of light, according to the dispatcher, who was perplexed at these events. The

A B-2 Spirit soars after a refueling mission over the Pacific Ocean on Tuesday, May 30, 2006. The B-2, from the 509th Bomb Wing at Whiteman Air Force Base, Mo., is part of a continuous bomber presence Photo: U.S. Air Force and Wikipedia

airport agreed to send future unexplained reports to Missouri MUFON.

After receiving more reports on October 31 I called the Lee's Summit Airport again. According to my contact at the airport the *KC Flight Team* left the airport at 6:45 PM and returned at 8:00 PM, flying over the Chiefs Stadium for Monday night football on October 31. This is a group from the *Experimental Aircraft Association* and there were seven planes flying in formation with this group.

The flight path included areas over Lee's Summit, Raytown, and Kansas City. The group flew in different formations, had a red and green flashing light on the wings, and bright flashing strobe lights that are typical of their night performances, but not typical for conventional aircraft. Debbie and I believe that this new type of night stunt flying will promote more UFO sighting reports in the coming years all over the U.S.

Media Coverage:

The UFO sightings in Kansas City were covered heavily in the newspapers, TV, Radio, and websites including the Lee's Summit Journal, KCTV 5, Fox 4, KSMO TV 62, KCMO 710 AM radio, Examiner.com (articles by Roger Marsh), Eye Witness Radio, Project White Paper, and more.

Many of the witnesses called the news media, sheriff offices, police, and airports to report their UFO sightings. Several people called me directly rather than file a report online. After the media calls started coming in to my office, I decided to write press releases in an effort to get the correct information to the public and stop rumors. Some reports claimed that all of the sightings were the experimental flying groups, while others made wild claims about the sightings that were misleading.

Some independent witnesses have submitted videos to You Tube that are not part of the MUFON investigation. Several of these reports also include data that is incorrect.

Linda Moulton Howe posted interviews on her website at www.earthfiles.com. Linda interviewed me and did a report on Coast to Coast AM with George Noory on November 22, 2011 about the October incidents as well as reports involving an alien encounter in April, 2011, and 125 balls of orange light in June of 2011.

While several of the October 4 reports can be attributed to the Flying Teams that flew over the Chief's Stadium, and perhaps some of the sightings were the B-2 Stealth Bomber, many cannot be explained as conventional or even experimental aircraft. Many of the UFO sightings occurred on different days and times, with the busiest days being October 3,4,5, 30, and 31.

Some of the remaining unexplained reports are following:

- 40-foot diameter orange fireball hovered approximately 60 feet above a vehicle for several minutes at an intersection in Lee's Summit, Missouri, then flew to the east at a high rate of speed.

- A woman and her teenage son saw a basketball size black sphere floating a few inches above Lee's Summit Road in Independence, Missouri at 10:30 PM on October 15 while stopped at a traffic light. The sphere traveled slowly from north to south uphill for approximately 80 feet, then simply disappeared.

- A man saw multiple small softball-sized white orbs in the woods behind his house in Lee's Summit, Missouri, which came out of the woods and went back. He watched the object for approximately 45 minutes, then they left headed east just on top of the tree line.

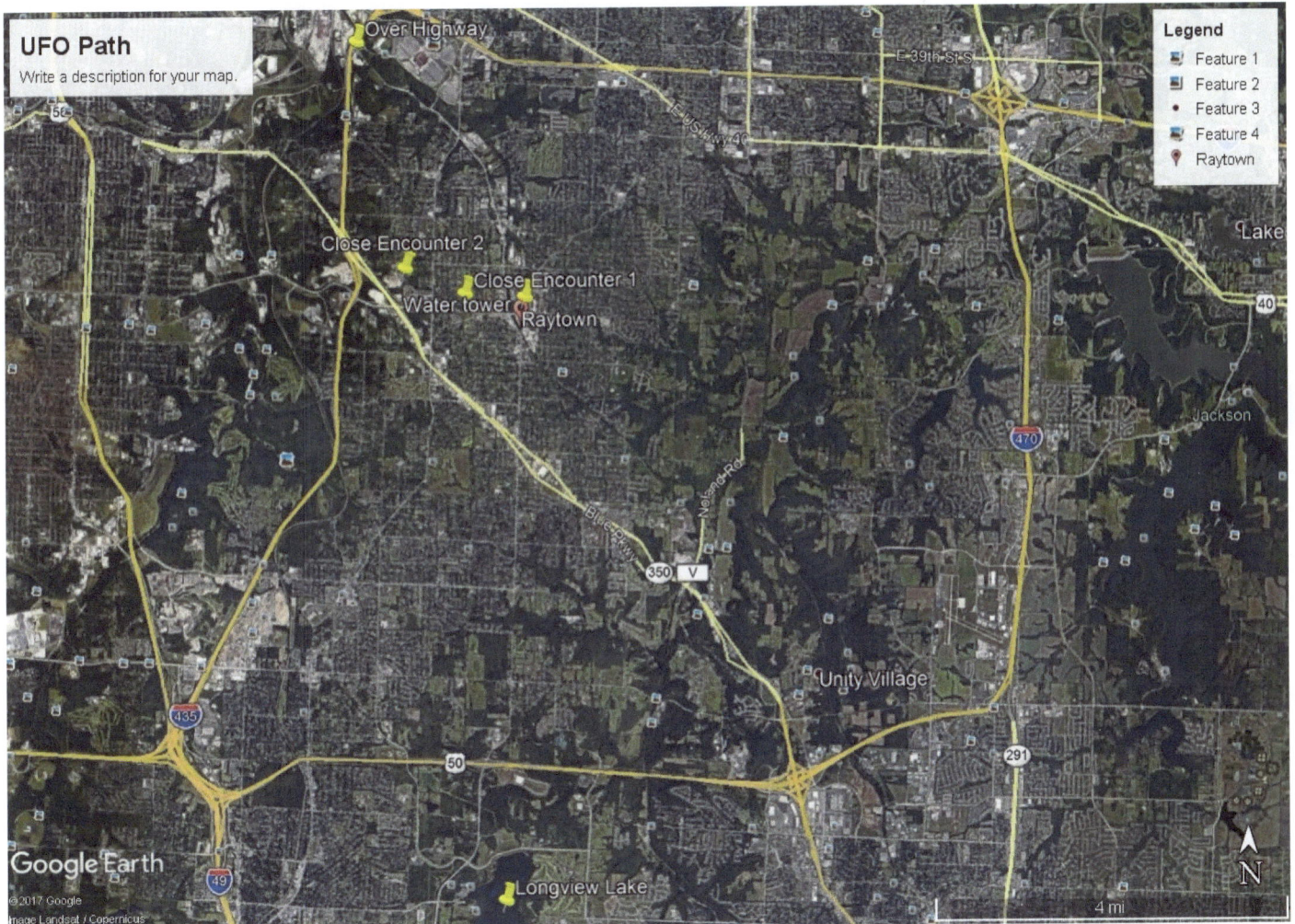

Map of the area showing flight path of large unidentified flying object on October 4, 2011

Photo: Google Earth

Timeline of events 10-4-2011

(This timeline is for the large disk-shaped craft only)

7:30 pm: Strange Gull Behavior at Longview Lake

8:00 pm: Large disc-shaped craft over 63rd and Raytown Road over the Raytown Water Tower

8:10 pm: Large craft over I-70 and 291 Highway

8:15 pm: Large disc shaped craft over 63rd and Woodson Road at the Apple Market (Close encounter)

8:30 pm: Large disc-shaped morphing craft over 63rd and I-435 Highway (Close encounter)

10:50 pm: Diamond-shaped craft over I-70 highway (morphed?)

11:00 pm: Large disc-shaped craft over I-435 and 152 Hwy

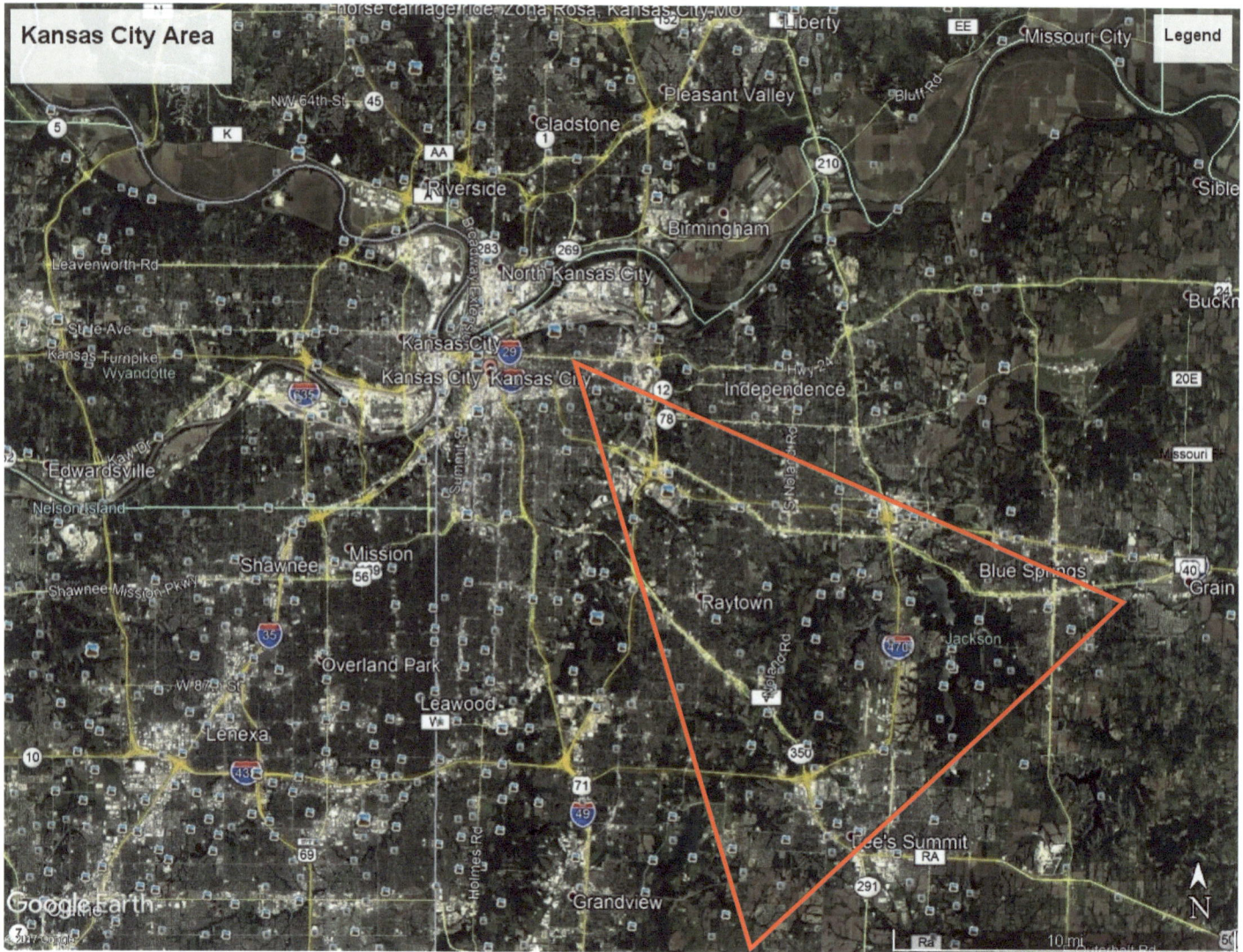

Image of the Greater Kansas City area and surrounding suburbs of Independence, Raytown, Blue Springs and Lee's Summit, where most of the sightings occurred.

Graphic: Google Earth

Links to News Reports and Videos About UFO Sightings in the Greater KC Area
2011-2012

JCMB News: http://www.kansas-city-news.pro/2011/09/ufo-sighting-in-kansas-city-2011.html

https://youtu.be/TMaQk3WkVDY: My own You Tube Video of a UFO over Independence July 4, **2012**

https://youtu.be/69ZNAICZyFQ: Mass UFO Sighting in Lee's Summit, MO

https://youtu.be/KJQ_ri5B0Qs: Video of UFOs October, 2011

www.youtu.be/N_bfrc5tPNI: KCTV 5 Report of Lights over Blue Springs and my investigation of these events.

www.youtu.be/aZc3jM_xxZ4: KCTV 5 Report of Lights over Blue Springs May 2012—object dropped!

https://youtu.be/KqRknPAlo_w: Good UFO video January 2012 (not investigated)

https://youtu.be/a3IoTbs_soE: Fox 4 News Covers cow mutilation event January, 2012 (investigated by me later) This was not included in the book but could be related to the UFO sightings.

Also search YouTube for more videos.

Summary of Large Disk Craft Sightings:

As indicated on the map, the large unidentified craft first may have been at Longview Lake, then it moved to the Raytown water tower, made its way to the Apple Market, and then to 63rd and I-435 at the apartment complex where it spent quite some time circling a mining area, then approached witnesses at a nearby apartment complex and where a close encounter with alien entities occurred along with a possible abduction.

Then the craft moved north along I-435 and was seen above I-435 and I-70 highway, and finally at I435 and 152 highway.

This is NOT the path that the flying team took, therefore, it could not have been the planes in that group.

The object was seen by multiple witnesses, at close range, and at different locations at different times.

The object was obviously traveling through the area, but for what reason we don't know. This was not a conventional aircraft. It was silent—with no sound of engines, and had no wings, tails or rotors.

What it was or who was piloting the craft we'll probably never know, but one thing is for certain—this is one of the biggest UFO close encounter cases I've ever heard of and I'm privileged to have been a part of the investigation, which is still ongoing.

Note: Many of the witnesses continue to have sightings and encounters.

We continue on the following pages with reports beginning on October 31, 2011. Again, an experimental flying team was doing a demonstration over the Chiefs Stadium during a short time period. The reports involving the flying team were eliminated, and following are only a few of the more interesting sightings and encounters.

10-31-2011
Lee's Summit UFO Hypnotizes Large Crowd on Halloween Night

One of the strangest cases I received is that of a large UFO in Lee's Summit on the night of October 31. A Yoga teacher and her daughter-in-law and two grandchildren watched a large disc-shaped craft with multiple lights on the bottom moving slowly at low altitude at 291 Hwy and Todd George Road in Lee's Summit while sitting at a busy intersection. They watched the craft for several minutes as the stop light changed several times. They noticed other vehicles watching the craft as well. The witness took four cell phone videos of the craft and amazingly, the Flying Team was in the background of one of the videos.

This is the only known video that shows both the Flying Team and a UFO. The craft seemed to follow them as they drove on, speeding up as they drove, and slowing down when they came to a stop light. The compass on a cell phone spun around and would not work properly. The craft disappeared twice below buildings, then appeared over a neighborhood just as they got out of the car to allow the children to trick-or-treat on Halloween night.

The trip from the witness' house to their destination should have taken ten minutes, however, they arrived at their destination 1 1/2 hours later. The two women and two children believe that they experienced approximately one hour of missing time. The vehicle was cold inside when they arrived at their destination, which the witness thought was odd since she had the heater on and it should have been warm.

The craft hovered over them as it emitted a low hum. The witness said she thought the size of the object was at least 100' in diameter. 20 to 30 witnesses watched and talked about the UFO as they stood around the neighborhood. Amazingly, none of the witnesses were frightened and remained calm as they

Photo:

Copyright by anonymous

watched the close hovering object, which was at treetop level. The witness stated that she thought that the low-toned hum emitted from the craft may have been something created to calm everyone. As the witness watched the crowd, they seemed to be in some type of hypnotic state, which concerned her. The UFO then left very quickly, and everyone went about their business, with the only person in shock being the primary witness.

The trip from her house to their destination should have taken ten minutes, however, they arrived at their destination 1 1/2 hours later. The two women and two children believe that they experienced approximately one hour of missing time. The vehicle was cold inside when they arrived at their destination, which the witness thought was odd since she had the heater on and it should have been warm.

The witness told me that she would have several of the other witnesses contact me, however, they all stated that they did not wish to speak with anyone about it.

Large Object with Bright Lights in Lee's Summit
November 27, 2011
7:20 pm.

First Sighting—Police Officer

"I am a police officer and was on duty the night of November the 27th in Lees Summit, Missouri. I was at the intersection if 291 Highway and 150 Hwy when I noticed a round object with very bright white lights to my west. It would take a baseball held at arms length to cover the object. The white lights had a blue hue to them and all of the lights were on the bottom of the craft. The lights were strobing, and there was no sound whatsoever. It was a clear night with no cloud cover, and the object was very low, just above treetop level. I watched as it hovered, and filmed it with the dash cam on my patrol car for approximately 15 minutes, then finally turned the camera off because I didn't want to use up memory on the camera. I called dispatch, who got a good laugh out of it, but they wouldn't take me seriously.

I requested a copy of the video when I arrived at the station and will forward that to MUFON when I get it. The object then moved to the west slowly at first, then sped up very quickly and was out of sight in a couple of seconds. I stopped at Quick Trip and several families came up to me and asked if I saw the UFO. I assumed that they would call in a report, but according to dispatch there were no other reports received.

I am disturbed that something like this was not taken more seriously by my supervisor and looked into. I

searched the internet and found Missouri MUFON, and decided to file an official report. I am a former Marine, have been a police officer for six years, and have never seen anything like this before. I am familiar with conventional aircraft, and this was not any type of normal aircraft. I saw a similar object on October 31 in the same location but did not file a report about it. This object was closer than the other one. I called another officer to come out and he saw the object, too. We both want to remain anonymous at this time. I am interested in helping investigate UFOs in this area."

Note: During my investigation of this case, I called the police chief and asked if I could see the video of this event. We made an appointment for me to visit the station two days later. On the day I was scheduled to look at it I called to confirm and was told that the video had been lost.

7:30 p.m. Second Sighting of the Same Object

A woman and her mother were driving on Adams Dairy Parkway in Blue Springs near the I-70 entrance at 7:00 p.m. when they noticed a lot of very bright lights to their left (west) near Colbern Road flying low to the ground. The witnesses said they were close enough to know that the object was not a plane, but far enough to not be able to make out the shape clearly. It appeared to be circular.

The object was flying, then hovering, then flying in the opposite direction. As the witnesses watched the object the bright lights disappeared as the craft appeared to turn, and they believe that they were looking at the bottom of the craft when they saw all the bright white lights. After it turned, the lights were gone and blinking lights of different colors appeared to outline the edge of the craft. The woman's sister and her boyfriend were driving behind them and as we pulled up to the stoplight they were pointing at it also. As the group drove towards Lees Summit on 291 Highway south they again saw the object, then it disappeared.

Silver Spheres over Kansas City
January 16, 2012
5:15 pm

Michael Burke was leaving a restaurant at 85th and Wornall Road in Kansas City at 5:15 p.m. and was looking at some very strange egg-carton like clouds in the sky when something shiny caught his attention. Mike saw a cluster of approximately 25 perfectly round silver metallic spheres flying to the east of his position at 45-degrees off the horizon and at least 1,000 feet attitude. The sun was setting and shone off of these spheres.

The objects were in a large circular shotgun pattern, but as they moved away from Mike they got closer together and into a tight formation. They then moved into the strange clouds and disappeared from view. It would take a garbage can lid held at arm's length to cover these objects all at once at the point where he first saw them.

These objects moved silently in a type of wave, undulating through the sky. Michael said "The objects were definitely NOT birds. I have spent a lot of time in the wilderness and am a hunter, and these were not like snow geese, Canada geese, or any type of bird I've ever seen. I've also never seen conventional aircraft like this."

Michael was alone and did not have a camera with him, but from now on he plans to have a camera on his person at all times. Mike said "I am college educated, own and operate two businesses and am not prone to hallucinations." Mike said that he has never seen anything remotely like a UFO in his life, but that now he believes that there is other intelligence in the universe.

Triangular UFO Hovers Over Cell Tower
March 1, 2012
9:30 pm

My daughter was driving with her two boys and had just passed the Independence Power Plant at 21500 E. Truman Road when they noticed an object hovering right next to the top of a cell phone tower. They stopped the car, and watched in amazement as the disc-shaped object hovered silently for several minutes, then moved away slowly to the North and then out of sight. The three witnesses were shocked at what they saw, and said that they had never seen anything like it. There was nothing to indicate that it was a conventional aircraft of any type.

UFO, Abductions, and Communication
July 27, 2012
39th Street, Independence, MO
Anonymous

This is one of the strangest UFO sightings I've ever investigated. I received a call from a witness in Independence who I'll call Bob Smith. Bob is a retired Deputy Sheriff and retired IT technician and engineer living with his wife and two dogs in a home off of 39th Street. I've spent so much time on this ongoing case that now Bob and I are friends, and he's even spoken at our local MUFON meetings. I can say that he is definitely not delusional and is, in fact, one of the smartest people I know.

Bob, an avid amateur astronomer and sky watcher, noticed that a new object appeared in recent months which he hadn't seen before. The object appeared in the same place in the sky at dusk before any stars or planets could be seen. This occurred almost nightly. The object would then move slowly across the sky starting at approximately 45 degrees off the horizon, and ending at 45 degrees off the horizon in a straight horizontal line in a period of approximately one hour. After that the object moved behind trees in the distance and could no longer be seen looking from Bob's deck.

Bob had two sightings previous to this. In 2006 he saw a foggy area between his roof and the trees. He noticed that there were two basketball-shaped white lights chasing each other in a figure 8 or infinity pattern. A third light appeared with the others but abruptly veered north then east before rejoining the other two. This repeated three times. Bob said that one of the lights had searchlights that went above and below the objects.

In October of 2008 Bob was driving home on 39th Street with his wife in the passenger seat, when they noticed a large gold arrowhead with the sun behind it below the horizon. The arrowhead was stationary with the point towards the ground. A commercial airliner was headed toward the object and the witness felt that the pilots had to have seen it as it was twice the size of the aircraft. When he arrived home, he ran to get a camera. When the witness returned, the object was gone but the airliner was still in view. His wife does not recall the event, even though Bob said that she did see it at the time.

Bob noticed a very bright star in the southwest sky. After noticing that the object was moving and flashing blue, white, and red, he went to get his flashlight (0145 hrs.) to signal it to see if it would respond. It responded by fluttering like a leaf three times. This activity went on for three nights in a row and some of the time a second similar object appeared to the east. On the fourth night, Bob decided to aim his outdoor Infra-Red camera on his security system upwards and in the direction where he had been seeing the bright object(s).

On July 27, 2012 Smith locked the house and set his security system as usual and went to bed. What he didn't realize at the time was that his life would soon change forever, along with his perspective on unidentified flying objects.

Bob insists that he locked the back door of the house, set the alarm and went to bed. When he awoke the next day to let his dogs out one of them jumped up on the door causing it to fly open. The rear door has a dead bolt and opens to the outside. The alarm system is a professional system and is also monitored. The alarm never went off, and was still armed.

Bob was concerned as this indicated someone might have entered his home without alerting the residents, dogs, or the alarm company. After he reviewed his security footage, Bob found interference caused by his air conditioning while the camera recorded in IR mode, along with some unidentified objects that appeared in his back yard and trees. That is when it decided to file a report and contacted MUFON.

I visited Bob at his residence along with Missouri MUFON State Director Debbie Ziegelmeyer, who happened to be in town for other events. During my interview with Bob he stated that he woke up at 1:00 in the afternoon on July 28, which is highly unusual since he usually gets up at 7:00 a.m. He also had blood on his pillow and a nosebleed, which he dismissed at the time but later wondered if an abduction had occurred.

Smith said that the security system has a 200-foot motion sensitivity range and was aimed slightly WSW. We watched the footage on Bob's big screen television and there were three areas with lights that are not visible in a control photo taken by Debbie. The first area was what looked like a white blob, distant at 150 feet in the trees. It remained that shape for about two minutes. Within five minutes, the object emerged from behind part of the branches and took on the shape of a domed saucer hovering at an angle (see photo above).

The second light appeared in the lower right corner of the screen. It reminded Bob of a sea urchin as it was a big white blob with many rays coming off of it. The third light was located at the top center of the screen and visible on the lower portions of clouds. There was also some commercial jet traffic below the clouds that went beyond the horizon.

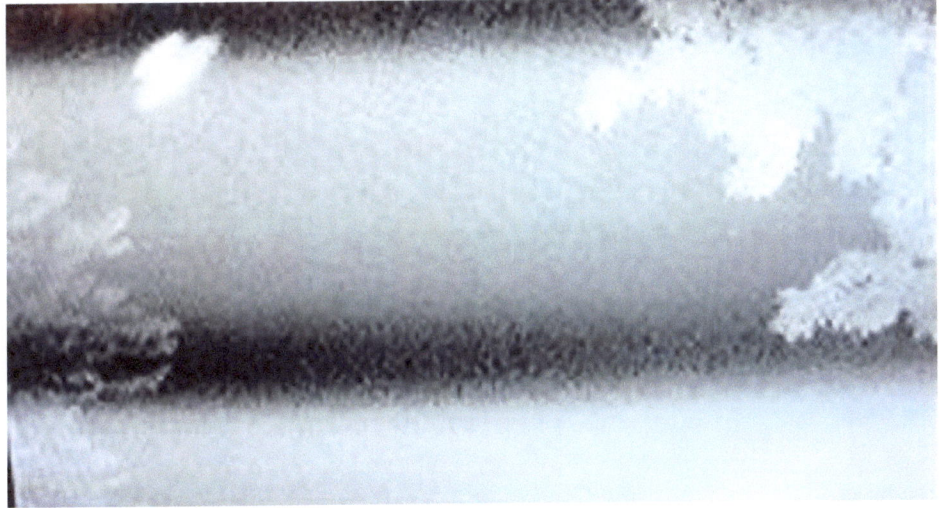

Screen shot of UFO from security camera video. Photo: Margie Kay Video: Anonymous

We watched this recording, which was approximately four hours in length, while Bob fast forwarded through much of it. The objects do appear very slowly, morph into other shapes, and even travel slowly across the screen. Debbie and I were at a loss for an explanation. I filmed this with my camera since Bob was not sure the video would be able to be retrieved off of the equipment at the time, but he did later.

The witness stated that the camera system is set up on a motion detector. The system did shut off during the recording even though there was always wind and branch movement that should have provided enough movement for the system to detect. It did this intermittently until dawn. The dogs were silent throughout the night.

Bob said that he wants his privacy to the extent that when asked if any of his neighbors have seen anything similar, he replied that he does not want to discuss it with them. He felt like family members he had talked with about it thought he was losing it. He said that the house immediately south of his is currently vacant so there would be no witnesses.

A few days after our visit Bob called to tell me that while he was on the back deck during the late afternoon someone may have shot a .22 at him which hit the barbeque grill and left a dent. He became concerned that someone was sending him a message not to talk about his sightings, however, we agreed that I would use an alias when discussing his case so he would have some anonymity.

Investigator L. J. visited the site a week after our visit. L. said "I made arrangements with the witness to return to the site after dark and did so at 2214 hrs. to photograph the rear of the house where the rear camera security system is located, just over the left shoulder of the owner. I parked a block away and walked up to the rear of the property on the sidewalk trying to keep silent as not to alert his dogs. His dogs began barking immediately upon my approach and I could see the TV was on.

The blinds opened and the witness came outside. I identified myself and was allowed to approach. The witness stated that he hoped his had not started something that he could not finish. When we concluded that no harm had come to anyone at this point he revealed that he shouldn't be afraid anyway as he was already a dead man. He stated that he was given a Stage IV Prostate Cancer diagnosis with expectancy of 5 years, 6 years ago. He still looked well. We joked that perhaps his medical interest was the common interest of his light visitors."

The below image is a control shot showing the view directly below the camera showing the night sky depicted in the owners security video.

L. and I examined the grounds to see where light sources may be that could have caused the images in question, but could find nothing. There is only one streetlight nearby, and it is clearly visible in the video. The other objects are dim and change shape. Even if the other objects were lens flares they would not change shape and move position. Additionally, Bob has other video from different nights that do not have these objects in it.

L. stopped where I thought searchlights might have been located directly down 39th Street at Walmart where the old Blue Ridge Mall had once been. He was trying to explain the third area of lights as he thought they might be advertising lights. He contacted the store manager who stated that they had no promotional event the night in question.

L. later contacted Jay at ATI searchlight Rental Company who told me that his company had no events that night and that I should contact Hollywood Lights in Springfield, Missouri. Jared stated that his company did not have any activity in the area that night either. This does not conclude that what was seen was not searchlights. It only suggests that if it were searchlights, the service was done by someone unknown or possibly an out of town contractor. The odd thing is that these lights were visible way past normal retail operating hours late in to the night and it is hard to conceive that an event would be held that late.

I remained in touch with Bob and he would periodically let me know about his sightings, which continued. I also stopped near this area on five separate occasions in an effort to see if the bright object in the sky still appeared, and did see it twice. I also observed something similar on at least 17 other occasions from different locations around the city including 87th and I -435 in Kansas City; Noland Road and 23rd Street, 23rd Street and Sterling, Truman Road near Lake City in Independence, Colbern Road and 291 Hwy in Lee's Summit; and I-70 and 7 Hwy in Blue Springs. Other witnesses reported similar objects in the same general area in eastern Jackson County and they have now become commonly known as the "Kansas City Lights."

One day, while Bob watered his front lawn a stranger approached him. The man, who was wearing a suit, came from nowhere with no vehicle in sight. He approached Bob on his front sidewalk and made some small talk for a minute then said "You know, Bob, a guy living near here was recently arrested for having child pornography on his computer." Bob, dumbfounded, watched as the man turned and walked away, down busy 39th Street and out of sight. Bob feels that this was a thinly-veiled threat.

Note: Other UFO experiencers have received similar threats.

After this incident it occurred to Bob to check his computer modem to see if anyone had been looking at his computer files and photos he kept of the object, as well as e-mails about the events. With his years of experience as an IT technician, Bob was able to locate this information on his computer and sure enough, there was not only one other person logged on to his computer, but TWO, and they both had U.S. Government e-mail addresses! Bob took screen shots of these logs, and gave them to me. Now we both knew that if people in government were interested in this case, it must be a big deal.

Bob lay low for a while after that incident, but several months later he put a blog up at www.Missouri-UFOs.blogspot.com in an effort to let people know what was going on here. Bob put his photos and videos on the site, which I saw, but less than two months later his blog simply disappeared. Attempts at getting an explanation from Blogger, which is a free blog application, were unsuccessful. At that point, he changed his e-mail address and we began to communicate in person, rather than via phone or e-mail.

Bob finally called me to say that he would like to attend a Kansas City MUFON meeting, which he did soon after and continues to do periodically. He told me that he had built a binary light board in order to send messages to the bright object in the sky to see if he could get a reaction. Bob brought this board to a meeting and showed everyone how he used this light board, as well as his million-candlepower flashlight to flash a message in binary code. His reasoning was that whoever was responsible for this object in the sky might be familiar with binary code. And "they" were.

Bob showed us a video of him flashing messages to the object, and the object responded by flaring up and repeating his code! At that point, I decided that we needed to get out to his house for a night investigation, which I did along with two other investigators that I invited to go with me.

We set up my video camera and high –powered binoculars, then asked Bob to do his thing. Bob set up his binary light board and we waited for a response. After about 20 minutes with no response, he decided to try the high-powered flashlight, and flashed three times at the object. To our utter amazement, the object flashed back at us! This occurred several times in a row, then stopped. I observed the object with the binoculars and saw that it had multiple bright colors and was flashing in approximate .33 second intervals. This was no planet or star—this was something that was under intelligent control.

The two other investigators and I decided to leave the house and go to another location and see if we could get a better view of the object. We found a parking lot and set up my camera, then flashed a light at the object—and it flared up in response again. Now the object was responding to us, not just the primary witness. We then observed the object moving behind trees and lost sight of it.

I kept in touch with the witness since and he has had ongoing sightings of what we now call "The Kansas City Lights." Many other people have reported the same type of light. In response to this, I started a Facebook page called "Kansas City Lights UFO Watch," which anyone can join.

Drawings of the orange orbs by the children

Multiple Witnesses in Missouri and Kansas See Orange Lights in Formation

June 16, 2012

Witnesses in the Briarcliff subdivision in Kansas City North in Missouri and at Wyandotte Lake in Kansas reported the same event. It is unusual to have witnesses from different locations view the same object, much less report it, so both Missouri and Kansas MUFON were glad to receive these reports from credible witnesses.

Debbie Ziegelmeyer, Missouri State Director of MUFON received the following e-mail though a link a the missourimufon.org website:

"Hello Debbie Ziegelmeyer:
The objects appeared to be emitting a fuzzy light from all over their surface. They reminded me of five Japanese lanterns with the light coming from within the object. The speed of each object was similar to what I would expect of a small airplane, like a Cessna 152. Maybe the most unusual part of this whole experience was that there was absolutely no noise coming from these objects. They certainly appeared to be close enough that had they been any aircraft that I know of, I would have heard a noise.

I have been a professional pilot for 58 years and most of my time was at very high altitudes where the weather was usually nice and clear. In all of my 20,000 plus hours in the air I have never seen anything that remotely resembled a UFO as described in most sightings ... until last Sunday night. While flying at high altitudes I have watched satellites go over, but never a UFO. I can remember maybe 30 years ago when pilots were reporting UFO sightings on a regular basis.

And in an e-mail to Debbie from a second witness: "I live in the Briarcliff area of KC North. Tonight my son and I witnessed 13 small reddish lights, orbs, traveling south in the night sky and then disappear in the trees. It started as groups of two or three, then a group of 4 and continued. My daughter saw 6 of them (she came out later) and my neighbor saw three (which he said 'vanished'). I would be very interested to know if you get anymore reports on these particular lights."

Debbie assigned the cases to me and I interviewed all four witnesses at the sighting locations. I took photos of the area and obtained drawings from each person. All of the drawings were similar.

The witness stated that there was no noise at all - since he was a pilot for 58 years he was shocked. Very surprised when the fifth one disappeared. Low altitude, neighbor thought she saw 13 objects but this could have been due to the objects going in a 360 degree turn four times. The path was similar to plane paths from KCI to the downtown Kansas City airport path. He first saw four objects evenly spaced, which went out of sight to the left (south). Then a lone light followed the exact same path and just disappeared due west of his location.

He called the tower chief - who was going to check with KCI radar and report back to him but has not to date. FAA General Council for this region is in the Federal Office building downtown. There are only two radars in this area, all radar from the area comes from KCI and Olathe Kansas City Sector, which covers large part of the center of the U.S. The witness has watched satellites go over his head at 50,000 ft. but never saw anything out of the ordinary in 20,000 hours of flight time. He wants complete confidentiality. The objects had rounded top and bottoms, smooth, deeper color at the ends than the center, more reddish white in the center, fuzzy outline, and horizontal lines or ridges in it. Flickered. 35 degrees off the horizon.

His position was off of 69 highway with a cliff on the right side - he was standing on top of that bluff looking west. You can see many miles away from that position.

NOTE!!!!! See the attached report #39444 from Kansas, which may be the same objects. If so, we have several witnesses from two different locations and if we find a third witness may be able to triangulate this.

Site Investigation
7/3/2012
1:20 pm – 2:30 pm

I met with the witness and his neighbor Mrs. Deborah C. and her two children of approximate ages 10 and 12 who live two doors down, and who was the original witness. I questioned them at each of their homes, then took photographs and obtained drawings.

Witness #1: Deborah C. (renter from Mr. K.) was sitting outside on her porch steps with her daughter when they both saw some objects through the trees. From her position trees cover most of the sky, but through one area she could see these objects. The witness called her son and he saw the objects, but was afraid and went back inside the house. Mrs. C. told me that she has seen other UFOs in the past but did not elaborate.

She said that she called Mr. K. who lives two doors down to have him look outside from his position to see what she was looking at in the sky. She described the objects as being round orbs or spheres, lit from within with a white light in the center and becoming red going outward from the center, deliberately moving in a straight-line path, one after the other in a group of four. She watched four more, then four more, then a final set of four followed by a single object that looked the same as the others and following the same path at 30 degrees off the horizon. The two children made drawings of what they saw:

Mr. K. walked across the street to the yard on his other property on a bluff overlooking the Missouri River, Kansas City, I-635 and for miles beyond. He saw four objects moving in a straight line from North to South at approximately 30 degrees off the horizon, then saw the objects move in a complete 360 degree circle, then off to the South in a straight line, following one another. A fifth object appeared further behind the rest and followed separately.

Both witnesses describe the objects as moving at the speed of a small plane.

From the other witness testimony, we have surmised that the first four objects probably circled four times, hence the first witness seeing what she thought was 13 objects, and this witness seeing 4 objects from a different perspective. Then they each saw the fifth object. With the amount of detail from both witnesses, I believe that these are not anything we can explain. This witness reiterates that he has been a skeptic his entire life, but is now quite certain that UFOs exist.

Investigation by Kansas Field Investigator Stan Seba:

Witness statement: "The assistant state director for MO MUFON recommended that I post this to the MUFON website. I hope you can tell us what we saw at Wyandotte Lake last night 6/16/2012. Five of us, age ranging from 59-79 were out on a pontoon boat last night about 10:00pm. Over the hill towards the east side of the lake above the tree line we saw 1 bright orange light about the size of a small street light rise up above the thick tree line. Then 3 more lights in a triangular formation came up. Then another triangular formation the same, rose up under the first one. The first single light disappeared or turned off.

The 2 triangle formations, with 3 lights each turned into 2 lights each. The 4 lights moved around and lined up to make a straight line. Then 1 single light came and 5 lights were in a straight line. The line of lights rotated counter clockwise and disappeared down into the trees. We all saw the lights about the same time. We were talking while it was happening. Saying things like look there, What is that? What are we seeing? I think we just felt curious, not scared or fearful. It all happened in a moderately slow, controlled way. After it happened we just wanted to see where the lights went. We expected to see more, but that was it. Afterwards everyone was talking about what happened. Mostly trying to explain it in a rational way. We even moved the boat to try and see more. It happened In the opposite direction on the Lake from the Legends. So we didn't think it related to activities there.

One person said, maybe they were helicopters. They were too close together for helicopters. In about 15 minutes 2 helicopters started circling the lake. They looked different than the lights we all saw, we all agreed on that. Wyandotte Lake is on the flight path of KCI. All night long we saw planes going and coming to and from KCI. The planes all looked and sounded like aircraft. This was silent, no noise. We saw something, we just don't know what we saw."

FI Stan Seba was assigned this report by Kansas State Director Steve Winans on June 18, 2012. FI Seba and SD Winans discussed this report via telephone on that date. We discussed specifically the location of the event, the witness's description of the objects described in the report and the history of the location of this report. Wyandotte County Lake is close to the Wolcott, Kansas area.

Wolcott, Kansas does have a history of unusual phenomena going back decades. There are reports within the MUFON CMS from that specific area. The most difficult thing will be determining the exact location/coordinates of the pontoon boat that the witnesses were on when they saw the object that was reported in the CMS. Both FI Seba and SD Winans believe that this data is important to the details of the report filed by the primary witness.

The area surrounding the Wyandotte County Lake is currently in economic development. It is very close to I-70 and I-435 highways, the Kansas Speedway, the Legends outlet and shopping mall and various new housing developments. The Wyandotte County Lake is also a landmark that many commercial pilots use to line up for landings at the Kansas City International Airport. (MCI) It is the opinion of FI Seba that these are all important points to keep in mind during the investigation of this report.

Interview of Witness Regarding Unknown Object:

June 21, 2012: According to the primary witness, she and four other relatives had taken a pontoon boat for use at Wyandotte County Lake. The primary witness and her relatives had spent the evening boating around the lake and admiring the countryside. According to primary witness the weather was clear for the most part. There were clouds in the distant eastern sky. Later towards evening the primary witness and her relatives noticed flashes of heat lightning in the distant eastern horizon.

At approximately 2200 hours, the primary witness and her relatives noticed to the southeast rising above a hill, three reddish orange lights in a triangular formation in the sky. At this time the primary witness and her four relatives were on the boat as it's bow was pointed southwest. Everyone on the boat immediately noticed the triangular formation of the three lights. One of the primary witness' relatives quickly maneuvered the boat to where the bow was pointing south. As everyone the boat was watching this formation of lights, a second formation and then a third formation of three reddish orange lights also arose from behind the same hill as the first formation.

At this time all five people aboard the boat were trying to figure out what they were seeing. As the primary witness and her four relatives were watching these two formations of lights, one by one, the formations pivoted (according to the primary witness) then realigned into a straight line formation and descended behind the hill from which they arose. At this time the primary witness in her relatives attempted to maneuver the boat south towards the geographic center of Wyandotte County Lake.

At no time did any of these formations of lights make any noise. Prior to the sightings the primary witness and her four relatives had seen and heard air traffic north of Wyandotte County Lake headed towards Kansas City International Airport. It struck the primary witness and all four relatives as strange that these three formations of lights made no noise whatsoever. According to the primary witness the duration of the entire event was approximately 5 minutes. The primary witness stated that all the lights in the formations were perfectly rounded shape, and not flickering or wavering at any time during the sighting. The primary witness did tell FI Seba that as the lights descended

behind the hill it was as if the lights were "turned off" as they went down behind the hill.

According to the primary witness about fifteen to twenty minutes after the formations of lights disappeared behind the hill she and her four relatives noticed there were two helicopters flying into the area from the south. The primary witness stated that two helicopters were circling the area where the three formations of lights disappeared. To date neither the primary witness nor her four relatives are able to discern what types of helicopters they saw after the three formations flights disappeared. According to the primary witness she and her relatives were unable to see any markings on the two helicopters. The primary witness did state that she saw the helicopter identifier lights and they were clearly able to hear the sounds of the main rotors of both helicopters.

The primary witness told FI Seba that while they were watching the helicopters they did move the boat farther south on Wyandotte Lake in order to see if they could get a better view of either the formation of lights or the helicopters that were flying in the area.

The primary witness told FI Seba that she and the other four witnesses at the time were upset by the sighting. Specifically they were all upset because they were unable to determine what they had just witnessed. According to primary witness they did briefly argue over whether or not what they had just seen was a natural, man-made, or extraterrestrial/extra dimensional phenomena. All witnesses are in agreement as to what they saw, but, they do not agree as to what these objects actually were. The primary witness did tell FI Seba that three of her relatives refuse to speak openly about this event.

The primary witness did volunteer to FI Seba that her three relatives' reasons for not discussing this openly or publicly has to do with their current professions. It is their belief that if their names are associated with this event it would be detrimental to their professional reputations. FI Seba did make it clear to the primary witness that no names would be used at any time within his reports and that he would honor all requests from all witnesses remain anonymous. FI Seba also made it very clear to the primary witness that he never uses or reveals the identity of anyone in his reports without their express and explicit permission.

Interview of Witness Regarding Post Sighting Anomalies:

According to the primary witness who was only willing to speak for herself at the time of the interview, she did experience severe nausea and a severe headache almost exactly 24 hours after her experience. In

Kansas City International Airport is located 9.5 miles North by Northeast from Wyandotte County Lake. Bearing 20.84° from true North.

Earthquake Activity:

No unusual earthquakes or anomalous seismic activity was recorded or reported at the time of this event.

Sighting Location Map:

Wyandotte County Lake and Briarcliff.
Below: Google Earth image of the area

after we saw these objects I must've thrown up at least 10 times. I have never in my life been so sick to my stomach as I was the night after I saw these objects." The primary witness did state to FI Seba that there was the possibility she may have contracted some form of stomach flu or temporary bug. After a good night sleep the primary witness awoke in the morning of June 18, 2012 with no ill effects, headaches, or feelings of nausea.

Besides this minor sickness primary witness did volunteer to FI Seba that she and her four relatives are a little "obsessed" to find out exactly what they did witness on the night of June 16, 2012. To varying degrees primary witness and her four relatives have done some research on their own and they have revisited the site where they witnessed this event at least once as a group.

The primary witness made it very clear to FI Seba that everyone involved really wants to know exactly what they saw. As a group they're very confused as to why they saw these objects.

Investigation of Other Witnesses:
June 25, 2012:
The primary witness arranged for FI Seba to speak with these two witnesses. The first witness FI Seba spoke to during the phone call was a relative of the primary witness- I will call her witness number two. Witness number two told FI Seba the same story of the event as did the primary witness who filed this report. Witness number two is curious and wants to know exactly what she and the others witnessed on June 16, 2012 while at Wyandotte County Lake. When asked about Post Sighting Anomalies, witness number two stated that she had no psychological or physiological effects from during or after the sighting. The only exception being that she has been a little obsessed in trying to find out exactly what she saw along with the four other witnesses.

The second witness FI Seba spoke to during the phone call was also a relative of the primary witness. I will call him witness number three.

Witness number three recounted to FI Seba the same story as the primary witness reported. The only exception to the reported account was that witness number three believed that the objects were much farther away from what the other witnesses have described to FI Seba. When asked about Post Sighting Anomalies, witness number three did state that he felt very excited when he saw these objects. Witness number three also stated that he has had at least one dream about these objects since the reported event. FI Seba asked him if there were any differences in his dream from the actual event, according to witness number three the dream was just like a replay of what transpired in the initial report. Witness number three went on to state that he believes that there is life in our galaxy that does visit the planet Earth.

Witness number three relayed to FI Seba that he is Brazilian and has a very firm belief in psychic phenomena. He also stated to FI Seba that throughout his life he has experienced clairvoyance- specifically the ability to see into the future with great accuracy. As he was growing up in Brazil, many people in the town where he lived would come to him and asked him about future events. Due to the accuracy of his predictions, many of his neighbors in Brazil told him he was a "medium." FI Seba asked witness number three if you'd ever seen unusual lights or UFOs in his past. Witness number three stated to FI Seba that the events on June 16, 2012 at Wyandotte County Lake was the first time he'd ever seen anything unusual in the sky. According to witness number three to this date he's never had his psychic abilities tested.

Nearby Military Bases:

Fort Leavenworth, Kansas
Ft. Leavenworth Military Reservation and Sherman Airfield is located 14.5 miles North by Northwest from Wyandotte County Lake. Bearing 326.14° from true North.

Forbes Air Force Base, Airfield, Topeka Kansas
Forbes Airfield is located 50.0 miles west by Southwest from Wyandotte County Lake. Bearing 252.75° from

true North.

Whiteman Air Force Base, Missouri
Whiteman Air Force Base is located 72.5 miles East by Southeast from Wyandotte County Lake. Bearing 114.54° from true North.

Pilot Reports UFO Sighting
7/30/2012

This witness called me directly to report a UFO sighting the day after the event. The witness wishes to remain anonymous. The witness stated "I am the pilot of a regularly scheduled commercial aircraft flight from St. Louis to Kansas City and was 25 minutes from MCI when my co-pilot said he saw an unidentified object to the right of the aircraft that appeared to be following us. It was not yet in my range of view, so I asked him to confirm that and the object remained. I radioed the tower and asked for radar confirmation but there was none.

Then the object came into my view to the right front of the aircraft. The object was a silver/gray color and had three blue/white lights on the bottom and one on the top. It was a disc-shaped craft with a low dome on top. It was approximately 300 feet from our aircraft and NNE of our position. Then the object suddenly moved underneath us and to the left, and continued at the our same rate of speed.

I discussed this with my co-pilot and we decided not to tell the tower what we saw because of what happens to pilots when reports like this are made. The tower then asked us to confirm an unidentified object and I said it was no longer in view. I was concerned, however, that it might be a problem. The object stayed with us for approximately 22 minutes. When I decreased aircraft speed for approach, the object slowed as well, then went directly south of our position at a very high rate of speed - faster than any aircraft I've ever seen.

I was a military pilot for 14 years and know what type of aircraft we have. It was definitely NOT a man-made aircraft. Some pilots have lost their jobs when they report a UFO, and I would like to keep mine so am not providing my name or contact information. Some of the passengers were discussing this as they left the plane but we did not discuss it with them. Other crew members saw it as well, but did not report it. I just wanted to tell someone about it."

2/11/2013 addendum to report:

William Pucket from UFOs Northwest offered to assist with this case and I said I would appreciate the help since his group has aviation experience and connections with the air traffic control. They did a very thorough investigation and were able to obtain radar data and written transcripts of the pilot communication with the tower. There was one slight discrepancy in his report regarding the method in which the news media obtained information about the event - they got the info from www.mufon.org, not from a local website. Their conclusion is that this is a hoax, although they say in their report that there is missing data from the transcript (which to me is suspicious).

NARCAP website:

http://www.narcap.org/files/narcap_IS07_Final2.pdf

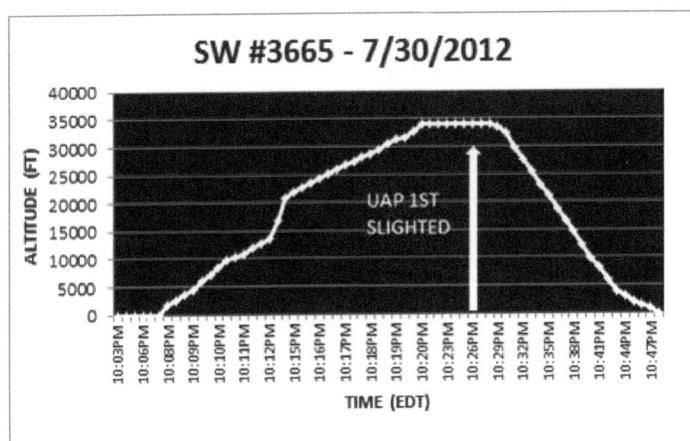

SW #3665 - 7/30/2012

Note: There are two reports from witnesses on the ground on the 30th and 31st: Case file# 41369 2012-07-31

1:05 pm

M.S. Longer than the jet it flew above. Cigar shape very visible until it disappeared after passing the jet at the highest altitude. I saw the jet smoke this silver grey huge object had no smoke trail. The sky was clear blue. I was at work on smoke break.

and

Case file #413352012-07-30 5:06AM

R. R.c Kansas City MO

See the KSHB TV coverage: **https://www.youtube.com/watch?v=R1mGVz75r4Q**

Cigar Shaped Object Follows Jet
July 31, 2012

A commercial airline pilot called my office to report the following: "It was July 31st 2012 a little after 1pm CST. I was on a cigarette break in the back of the building where I work. I work for the city ambulance service in this town. I was looking in the sky at a jet at the highest cruse level an airliner can fly when I noticed this grey/silver long cigar like object about 50ft or more above it coming from behind the jet. It continued to fly over the jet, in front of it until it disappeared in a small thin cloud. I waited for it to reappear after it existed the cloud but it never did.

This day the sky was clear blue with very few thin clouds, not to mention we had triple digit temperatures. I was not hallucinating from the heat as my co-workers joked about when I told them. I have never witnessed a UFO in my life but that day I know I saw one.

I can't get this thing out of my head now. And the chilling thing about it is that night channel 5 KSHB reported a story from a pilot who said one trailed their plane as they landed at KCI airport the day before on Monday July 30th.

 The jet looked like a reflective dot with a very long smoke trail. This object stretched five times that of the jet if not more at that altitude, it was amazing. I am no-longer a doubter, I know what I saw and it was not man made. I wish I had had a camera but I did not I just hope I was not the only one who saw it. "

Balls of Light at Sunset—similar to some balls of light seen by witnesses. Photo: Fotolia.com File: #122713596 | Author: Balate Dorin

The Unexplained Balls of Light

There have been literally hundreds of reports of strange unexplained balls of light or orbs in the greater Kansas City area since 1969. A large concentration of these occurred during the years of 2011 to 2014, however, these objects continue to be viewed and reported not only in the KC area but across the state of Missouri and the United States. The colors most often reported are white, blue, red, orange, and black. See more about the unexplained lights later in this book. Following are some of the more significant sighting reports I've investigated: (not in chronological order)

Light Balls in Trees
October 3, 2013
11:45 pm

A Lee's Summit man watched five white balls of bright light the size of softballs dance around in the woods behind his house as he waited for his dog, who wanted to go out. The witness stood on his deck as he watched, and described the lights as "If they were playing." One of the lights moved into the yard area approximately 25 feet from his position and six feet above the ground. The ball of light moved back into the trees and the dog followed. The dog then came running back into the house and acted strangely the rest of the evening, refusing to come out from underneath the dining room table. The witness watched these balls of light for approximately 45 minutes, until they moved away to the east along the tree line and out of sight.

Huge Orange Fireball in Independence
January 27, 2014
9:09 pm

I had my own sighting and reported it to multiple sites: "I am the Assistant State Director for Missouri MUFON and had my own sighting last night. My husband and I were leaving Bass Pro Shops in Independence, Missouri near 291 and 40 Highway when something caught my attention in the northwest sky at approximately 40 degrees from the horizon. I watched a very large flaming glowing orange fireball cross the sky from the northwest to the southwest in a straight-line path for six to eight seconds. It was perpendicular to the ground and never changed course. The object left no trail whatsoever and appeared to come to an abrupt stop, then fade out like a light bulb in less than two seconds.

I did not hear a sound but did feel a strange vibration, like a shock wave. There was nothing else in the vicinity (planes or helicopters) and the sky was clear. The object looked like a complete sphere that was made out of orange flames. The size was almost the size of a walnut held at arm's length. It appeared to be about 500-1,000 feet away from me but it is impossible to tell distance. Unfortunately, my husband did not see it as he was getting into the vehicle at the time and I did not have time to get my phone out to take a photo."

Ball of Light at Watkin's Mill
June 30, 2012
9:30 pm

Witness statement: "Our family was camping at Watkins Woolen Mill State Park and it was about to rain, so we were rushing to put the rain guard on our tent. About halfway through, my wife asks "What is that up in the sky?" I am a firm disbeliever of alien life visiting this planet and UFOs attributed to them, so I looked-up and saw something I could not explain.

It was a slow moving ball of light that flickered as it moved. It was completely silent. About 2 minutes into viewing it, the light shut off and then turned back on. It was slowly descending into the horizon, which was trees, when it turned to the left of the horizon very quickly, went really fast, there was a flash and then it disappeared. I could not explain the object, so I told my wife it might have been an experimental govt aircraft, a meteor, or something else scientific.

I had this eerie feeling all night afterwards and in the middle of the night my wife and the other 3 women in our party walked-up to go to the bathroom, leaving me

Historic UFO Sightings in the Kansas City Area

Aliens and Craft in Kansas City
1956

As stated by the witness "I have never told this story to an agency before, but here it is as I remember it. In 1955 our family moved to the county outside the city limits south of Kansas City to a new home development that was being built. So there was new home construction. This is important as you will see. I was 5 years old when we moved. Either that year or the following year, my grandparents came to visit us. It was a beautiful sunny summer day. The sky was blue with a few white puffy clouds.

Grandfather and I took a walk south on Holly Street from 97th to 99th street to a ravine a few yard south of 99th street. We looked down into the ravine and there was a shiny, silvery disc-shaped object that seemed to be floating a few feet above the ground making a humming sound. It seemed as if the object was made of three pieces, a top (like a dinner plate turned upside down), a ribbed center section that seemed to be spinning independently of the top and bottom plates, and a bottom (like a dinner plate turned right side up). But I really can't be sure if it was just the center section or the whole thing that was spinning.

There seemed to be several small people there. I was not afraid, but I was very excited and happy to see them. I pointed at them and said something to Grandfather, but he seemed to be in a daze and paid no attention to me. The next thing I remember we were standing in front of my house a few blocks up the street. My Grandfather was still in a daze. I can remember trying to talk to him about it but he seemed to be unresponsive.

I was very tired after that and went to sleep on my parents bed. I remember thinking about that before I went to sleep wondering what had happened. Before I went to sleep, I remember thinking that I had seen some plumbing sticking up out of the ground from the construction. And that is what I told myself that it was. I did not remember all this until I was in my 40's running around the track. Then the whole thing came flooding back.

There were many other personal consequences I suffered as a result of this event. For example, when my parents took me to see the fairy princess, she was wearing a silvery dress with a wand. I was terrified and ran through the department store and hid inside a rack of clothes. I was terrified of needles and getting shots. At night, I surrounded myself with stuffed animals for protection and when I looked at the window in my bedroom I thought about being on a flying carpet and flying through the sky.

I never understood any of this until that day on the track when everything came flooding back. After that, I had constant, nightly dreams of UFO's, abductions, being in spacecraft. seeing aliens. These dreams went on for years. About 7 years after that day on the track those dreams, thankfully, went away. I also have a lump in the crook of my right arm which I never understood till that day. It's still there. But that is a whole other story. I could make a drawing of the craft. That is crystal clear in my mind. But I have no way of putting it on my computer."

The witness told me by phone that she has since seen different types on non-human beings appear in her house, but she does not know who they are or what they want.

1969 UFO sighting at Lake City Army Ammunition Plant With Multiple Witnesses

Reported in June, 2014
(Witness does not wish to remain anonymous—this is his real name)

Lake City Army Ammunition Plant was established in 1940 and manufactures small arms cartridges, pyrotechnics, small caliber ammunition, and performs reliability testing of ammunition up to .50 caliber. It is housed on 3,935 acres with 458 buildings, 40 igloos, and storage capacity of 707,000 square feet. The plant generated large quantities of hazardous waste and is on the U.S. EPA National Priorities List. An interview with a former employee indicates that Nuclear materials are used on some specialty types of ammo.

I received a call from Raymond Griggs regarding his sighing of a large UFO over Lake City Army Ammunition Plant in 1969. The witness, who was told to keep quiet about the event, recently decided to come forward in order to let the public know that an unidentified object hovered and moved about the plant for two hours in the middle of the night and was witnessed by at least 13 people.

Raymond Griggs, age 22 at the time, was living in Independence, Missouri and working as a civilian contractor at Lake City. Lake City Army Ammunition Plant is a 3,935-acre U.S. government owned, contractor-operated facility located in eastern Independence near 7 Highway and 28 Highway. The plant manufactures ammunition for the military for the U.S. Army. In addition, Lake City performs small caliber ammunition stockpile reliability testing and has ammunition and weapon testing responsibilities as the NATO National and Regional Test Center.

Raymond worked the night shift from midnight to eight

Head stamp of a .50 caliber cartridge casing made at the plant in 1943. By $1LENCE D00600D - Own work, CC0, https://commons.wikimedia.org/w/index.php?8555293

O'clock A.M. in the printing shop. On a hot August night in 1969 the electricity for the entire plant and all buildings suddenly went out, so the crew of seven in his department went outside to cool down. That is when they noticed a UFO.

A bright chrome spherical object hovered just over a building which was located four buildings away from the men's position. The object was so bright it was blinding. The object then slowly moved over another building across the road and hovered, then continued to zig-zag back and forth across the road from building to building for approximately two hours. The seven men on the crew with Mr. Griggs talked amongst themselves while watching the UFO. At the same time, several jeeps with MP's were driving to and from the area very quickly, which is something the men had never witnessed before.

From their vantage point on the dock they could not see other people on the property at other buildings, but believe that everyone working there must have been outside since there were no fans running and the temperature was over 100 degrees that night.

The unidentified object then left the area by shooting

up into the sky at a 30 degree angle at an incredible speed. Raymond said that it was gone in ½ of a second, and that solidified the fact that the men were looking at a real UFO and not a military craft of any type. Just as the UFO exited, the electricity came back on.

The next morning, two men arrived at the office that no one working there had seen before. One was dressed in a gray suit and the other in black. One of the men said to everyone "Remember that you have all signed a security clearance and you did not see anything." The men then turned and left. They did not introduce themselves or show any identification. The message was clear, and Raymond Griggs kept his mouth shut for 45 years. But after seeing an episode of Hangar 1 in which Missouri and the Kansas City area were featured as a UFO Hot Spot he decided to come forward with the story.

I met Raymond Griggs son-in-law, Michael, at a MUFON meeting in Columbia on June 10, 2014. He told me that Raymond went public because he wants others from that night to come forward as there were over 1,200 men and women there that night that likely witnessed the event. Michael said that Raymond told him about this event only 1 1/2 months ago and he has known him for 30 years. Michael said "The way his father-in-law puts it is, 'Hey I am 69 - what are they going to do to me now?'" Michael has never seen a UFO himself, but after hearing about this is very interested in the topic.

After I submitted a press release about the incident, more witnesses came forward about similar events at the same location years later. These events occurred at or near Lake City in 2001, 2013, and 2014. In 2014 an employee arrived at work at approximately 7:00 a.m.

See more listings at http://thecid.com/ufo/k.htm

No one else was around at the time. He climbed the stairs inside his building, looked out a window, and was shocked to see a UFO hovering over a nearby building. The UFO was silver in color and saucer-shaped, and measured approximately 30' across. The witness watched it for a couple of minutes, then the object shot straight up and out of sight in less than a second, leaving the witness dumbfounded.

Many websites picked up the story, as well as local and national radio talk shows including Spirt Talk with Victor Furhman and Chris George, Nightwatch Radio with Todd Sheets, and Open Minds UFO Radio with Alejandro Rojas.

The 1970's and 1980sUFO Flaps

In doing research on historic UFO sightings I found several newspaper articles and photos about a number of huge unidentified craft that appeared in the skies over Kansas City in the mid-1970's when there were multiple sightings across the country, and again in 1987–1989. At that time no one was connected to the internet, so news traveled slowly. Apparently in 1987, a number of people witnessed unidentified flying objects over downtown Kansas City and the surrounding areas, many during daylight hours. Some people got photographs of the objects which I saw on the internet several years ago, but since then I have been unable to locate these photos again. Some were quite spectacular, showing extreme detail.

There was a lot of talk about a UFO Flap at that time, and I've spoken to people recently who remember the event was covered on the local news, however, I've been unable to obtain further information about this. I mention this event only because so many credible people have talked about it. Perhaps we can find some more evidence in the future. Following are a few of the more interesting reports from that time period:

Hovering Dark Object

Kansas City, Kansas
1987

Witness statement: "Dark object, shaped like Chinook helicopter but no sound, with red flashing light on left side, another on opposite end that stayed lit. Went out of sight behind hill, vanished after they topped hill. A hovering object was observed. One dark oblong object was observed by three witnesses (Daniels). No sound was heard."

Source: Walters, Edward UFOs are Real: Here's the Proof, Avon Books, New York, 1997

Craft Hovers over Car

Kansas City, Missouri
December, 1989

Witness statement: "My family and I were searching for the Metro North Mall. We are from Saint Joseph and were not familiar with this area. It was sometime around Christmas. We were shopping for presents. We ended up on a road that lead us somewhere around a open field. It was night and we saw headlights approaching us. My father did not know how to get back on the highway, so he was going to wave the other car down and have them tell us directions. My father came to a stop and waited for the other car to pull up. Except it stopped about 25 feet in front of us. That is when we notice that the head lights seemed to be floating in the sky. After a few seconds my father pulled a little further up, and the lights raised about 15 feet in the air. My step-mother freaked out at that time and opened her window. She thought that it may of been a helicopter(I have since lived by a hospital that life flights out and know better). There was absolutely no sound what so ever. Not even a car motor. It remained there for about a minute and then it raised a little more. Then, suddenly it shot kind of diagonal into the sky. And eventually straight up and out of sight. I would like to put that the speed that it traveled was amazing. Not helicopter like at all. I do not know if it was from this world or not but I do know what I saw."

Source: National UFO Reporting Center

Three Objects Seen in Triangle Formation

March 20, 1998
10:50 pm

Witness Statement: "I observed a bright pulsing (red, green, and white) light with several smaller white lights around it, while riding in the car with my sister. When the whole formation started twinkling and moved almost instantaneously. We pulled over to the side of the road to observe when the smaller lights started twinkling and turning red and darting away from the formation, with really small lights coming from the red lights. One of the small lights came within 20ft. of us just above the tree line. It was very small, the size of a headlight, no sound, no rays of light coming from it. When it came to the closest point to us we felt a rush of fear come over us then it was gone. We continued to watch the larger pulsing light move NE in intervals until it was out of sight."

Source: National UFO Reporting Center

Tumbling White Object

Three Massive Black Triangular Craft

Riverside, Missouri
April 20, 1989

Witness statement: "It was about 9:30PM and I had stepped out onto the porch of our third floor apartment. I looked to the south and saw a very unusual aircraft going east. It passed just below a full moon. Although I was 14 at the time I had a very deep interest in aircraft and new this was not normal as it did not have the standard FAA lighting.

Once the craft was out of view I went inside and told my mother what I had seen. She shortly left for work. About 20 minutes later I return to the porch. After standing there for a few minutes I look up to see three of the same aircraft flying in diamond formation. I was in awe. The aircrafts were at roughly 10-15,000 ft.

Now I was used to seeing airliners coming into KCI/ (MCI) all the time. These craft were flying above the normal approach altitude but below the normal transcontinental altitude. The only thing was even at their altitude they appeared at least three times larger than a conventional airliner at the approach altitude.

They were massive. They moved slowly, taking at least two minutes to leave my field of vision that was fairly confined as I was in an apartment complex. The other thing that I noted is that at no time did I ever hear engines of any kind. One detail that I wish to mention is the circle in the middle of these triangles. When I saw the first one at a profile, this circle appeared to be more of a dome on the bottom of the craft. Then later looking up at them this dome had white light but also appeared to have swirling colors within that white light.

At this point I went back inside. I then placed a phone call to the control tower at KCI/ (MCI). I asked the man who answered if there had been any aircraft fly formation over the Kansas City area anytime in the last 48 hours. He said that he didn't know the answer but if I could hold he knew who would. After being on hold for about five minutes he returned and informed me that indeed there had been no aircraft flying formation in the area in the last 48 hours. I then told him that we had a problem because I just saw three large aircraft less than 15 minutes ago in formation. I explained what I had seen. He then asked if I would like to report a UFO. I said that I did. He said that I would

need to hold as he put in a conference call to the, "United States UFO Reporting Agency". After being on hold for about five minutes the man from KCI introduced another guy that was on one of the worst phone connections I have ever heard. It was distorted, scratchy, a lot of line noise. I gave the man a detailed account of what I saw. He took my name and number and said that they would be in touch. I never heard back from anyone."

This case was reported to MUFON in 2006 and was investigated by me. I called the witness to get more information:

I spoke to the witness on 6-29-06 and he stated that the objects each looked slightly smaller than a C5A transport plane but were higher up. They were each the size of a nickel held at arm's length. There was no sound. The witness has seen eight UFOs in his lifetime, one with another witness. These sightings have prompted an interest to be a MUFON Field Investigator.

There are many more reports during the 1987-1989 period in and around Kansas City. To see more witness accounts visit the UFO DNA website at http:// thecid.com/ufo/k.htm, The National UFO Reporting Center at www.nuforc.org, or the International MUFON website at www.mufon.com

Alien Appears in House
August 12, 1989
Independence, MO
8.00 pm (approximate)

On the evening of August 12, 1989 my husband and I went out to dinner with friends and left my two daughters, age 12 and 10 at home. The girls had two of their girlfriends over to visit.

We had just finished putting up drywall on a new addition of the house next to the kitchen, which was about 20 feet away from the kitchen. Suddenly, the girls noticed movement on the wall next to the stairs. The four of them saw a large face appear on the wall. It measured approximately 24" high. It was the face of a male entity wearing an M-shaped helmet. The helmet came to a point in the middle of the forehead, and down the sides in front of the ears. The two girls who were visiting ran out of the house and went home, and never came back to the house again. My daughters were used to this type of thing and just stared at the face, wondering what it was.

I had been visited by the exact same face twice before – both in 1986, but I hadn't told my children about the incidents. (More information about these contacts in my upcoming autobiography). Now I wonder if this same being didn't also give information to my daughters as he did me, but they just don't remember the communication.

Large Craft Hovers Over House in Busy Neighborhood
Fall of 1989
Kansas City, MO

This witness called me directly. She is currently in her 60's but remembers this event "As if it were yesterday." The witness said that she was driving home from work one evening and at approximately 5:30 pm in the fall of 1989 and as she approached her house, which was located on a hill in a neighborhood with many houses, she saw a large object directly above her house. There was no one else around or outside at the time.

The witness stopped her vehicle because she was stunned by what she saw. A large disc-shaped silver metallic craft with a dome shape on top and rectangular shape windows on the sides was hovering above the house. There was no sound whatsoever, but she noticed a slight vibration. She could see movement in the windows, but nothing was clear. The witness looked around to see if anyone else saw the object, but oddly, there was no one else in the area at that time, which was unusual. The witness said that at that time of day there is usually a lot of traffic and people around.

She watched the object for approximately 10 minutes, then it ascended vertically at a high rate of speed and was out of sight. She stuck her head outside of the car window and looked up but the object was gone. The witness said that she was in shock, and afraid to go into the house by herself, so she waited until her son got home and they went inside together. He checked the house thoroughly but there was nothing out of place.

This incident made her so uncomfortable that she sold the house two months later and moved to another city. She called me after seeing the UFO sighting reports on a TV program in 2013.

Tiny Extraterrestrials?

As bizarre as this sounds, it may be true. Several witnesses, including myself, have seen tiny UFO's and small orbs with what look like extraterrestrials inside of them. I'll share three of these cases for the reader's enjoyment.

Case#1: Small UFO's Inside a House

One of my family members asked me to investigate his house for paranormal activity, ghost sightings, and tiny UFO's. This house was built in 1910 over an older foundation in a very old area of Independence which was bustling in the 1800's.

This all started about seven years ago. One evening, my relative and his spouse had just gone to bed when they both saw a small 10"—12" diameter silver disk with windows on it come through a wall and go to each corner of the room, hover, then flash as if it were scanning the room and taking pictures. It then moved over the bed and at that point, the two saw tiny people or aliens inside the windows of the craft. The two people just stared, frozen, while the object hovered. The craft then turned and went back through the wall where it came from.

A year later the two were shocked to see a bi-plane type craft which was approximately 8" - 10" in width come through the wall and fly around the room, then go back through the wall.

I believe my relative and his spouse—they are both practical, intelligent people who are not looking for attention. They just want to know what is going on and unfortunately, I have no answers. I do know that their house is a hot bed of paranormal activity and I get strong readings of Leylines when I dowse there.

Case #2: Pilot of a Craft in an Orb?

I investigated a site for a friend and her neighbor in a location outside of Kansas City. They had been seeing balls of light come over a hill and tree line every night at dusk, and were experiencing a lot of unexplained noises, lights, and sounds inside and outside the house. I spent the evening at the site and set up cameras. As soon as the sun went down, we all watched multiple glowing orbs appear over the tree line and slowly move down into the trees in front of the house.

I took photographs of these transparent softly glowing white-blue orbs and in two of them there are what appear to be small alien pilots sitting at controls. The orbs are no bigger than 6" in diameter.

Case #3: Another Tiny Pilot?

This location is right on the old Santa Fe Trail in Raytown, Missouri. The witness and his wife began seeing UFO's a few years ago in many locations, but mostly at their house and in the yard. The primary witness even set up an elaborate security camera system with night vision, and he captured images of extraterrestrials standing right next to him in his yard. At the time, he could not see them with the naked eye, but could feel a presence.

Debbie Ziegelmeyer and I went to the location to investigate and set up several cameras. She and I and the primary witness observed several unidentified craft in the skies above. At one point, we all felt as if we were being watched by something close by.

Debbie and I took photos, and in some there are several bright and dull orbs that appeared on the night vision camera and regular cameras. In one picture that Debbie took, there is a tiny extraterrestrial sitting at a control panel inside the transparent orb. The orb was approximately 6" in diameter.

Is it possible that there are tiny extraterrestrials? Or can they shrink to a smaller size in order to do reconnaissance? These are questions that need answers.

The Infamous "Kansas City Lights" Ongoing since 2011

Blue light sunburst background. © rea_molko– Fotolia.com

The lights were first noticed by a Blue Springs Neighborhood in April of 2012 and reported to me by a reporter for a local TV station, and I called one of the witnesses. This event soon ballooned into an on-going investigation and sightings of these objects by multiple investigators since. These reports have been covered in the media and on YouTube.

The witnesses all saw strange objects doing bizarre things.

In May of 2012 the now famous "Blue Springs Lights" covered in a Hangar 1 show in which a TV news reporter and his camera operator, along with at least 15 others witnessed bizarre objects in the sky, sparked my interest in these bright objects in the skies that could not be explained.

The reporter, Dave Jordan, with KCTV 5 News, responded to the witnesses request for answers, and went to the Blue Springs neighborhood along with his camera operator. The two were expecting to see typical "lights in the sky" or planets, but after interviewing witnesses and watching the skies for a couple of hours, one of the objects became very bright and flashed multiple colors. Witnesses also noticed red and white laser-like beams shooting across the skies in all directions.

This object was northeast of their position, possibly over Independence. Suddenly, the object shot off to the east to the horizon and instantly back again to the same spot where it originally appeared at dusk! The camerawoman was so scared, she ran to the van and got in and refused to come out again, or return to the site.

The next evening I arrived at the site along with my friend and police officer Corey Pearce. We were setting up my high-powered binoculars and facing west when a bright laser-like beam of light shot across the sky just as I looked into the binoculars. It temporarily blinded me. Corey and others at the scene saw it, too.

Some of the witnesses said that was exactly what they had been seeing for two weeks.

We kept watching the skies as the sun went down and it got dark. Suddenly, two objects in the sky brightened and started flashing. I assumed they were both planets or bright stars and told the reporter that was what I thought they were, and that the flashing was due to atmospheric disturbance. However, when I checked Google Sky Map the planet Venus and the star Arcturus were not in the same positions. In fact, Venus was below the horizon and not yet visible.

To confirm what I was seeing I called Debbie Ziegelmeyer in St. Louis and asked her to take a look at Arcturus to see if it looked anything out of the norm. She said it did not. At that point, I became much more interested in what these two bright objects were that were each flashing three colors of light in a pattern.

At one point, everyone in the crowd became more interested in one of the objects, so I aimed my binoculars in that direction. I was shocked to see a silver colored grid over the light—something like a giant cage. Then I was even more shocked to see something rise up above the object on the right side, come down over the top of the object, then go back up and behind it.

The second object looked something like a transformer with a head, shoulders, and arms, but metallic and robotic looking. Just as I saw this I said "Oh my God," and everyone turned around, and Dave Jordan asked me what I saw. Thinking fast I said "I've never seen such bright lights before." I didn't want to say what I actually saw on TV for fear of causing panic.

I met with Dave Jordan and his cameraman at another location one evening to see if one of the lights would appear where I suspected it would—over Lake City Army Ammo Plant in Independence. Sure enough, a bright flashing object appeared where I thought it would.

Dave showed me the footage that they obtained with their good cameras and got my reaction. These interviews and film of one object are on YouTube.

What is not in any of the TV reports is that Dave said he was called into the station manager's office and was told that there would be no more coverage of the objects in the skies, and that this order was coming from "higher up." Dave is now working at a station in Florida.

These objects are blinding, larger than planets to the naked eye, have multiple colored lights that spin, and though high-powered binoculars appear to have a metallic structure around them or in them. I have witnessed these objects on numerous occasions and have been able to obtain film footage of them. Some even react to a 1 million candle-power flashlight and flash back at us!

I am of the opinion that these lights are actually structures of some type, and may be related to the Lake City Army Ammo Plant in Independence. That is the only location in this area where large bright spherical metallic objects have been seen close to the ground, just above buildings.

There are a lot of unanswered questions. What are the objects? Why are they here? Where do they come from – are they "ours" or do they come from elsewhere? If the objects are some type of new weapon from Lake City, that could explain why there are so many sightings in this area, and perhaps explain the laser-beam like lights.

What to Look for:

- Super-bright neon like multi-colored lights that flash at .33 second intervals

- Colors to look for are red-white-blue or red-white-green

    - Always three colors flashing

    - The lights spin from right to left

- The objects appear to have a metallic grid over them if viewed through high-powered binoculars

- The objects can appear in any direction, and remain stationary through out the night, or move with the stars, then just disappear

- Sometimes the objects drop to the ground suddenly, or shoot off horizontally for a long distance, then return to the exact spot where they started

- Sometimes flashing a high-powered flashlight will evoke a flash-back response

- These same objects have been reported in other areas of Missouri and other states.

    - Best viewed through high-powered binoculars, but they are visible to the naked eye in a dark location

- Sometimes there are two or three similar objects in the sky at the same time

    - The lights are visible year-round

Conclusion

My Grid Theory

The areas where most of the UFO's and balls of light are reported in the State of Missouri are along 39, 38, and 37 Degrees Latitude. This is something I took note of in 2011 while working on Sasquatch, animal mutilation, paranormal and UFO investigations. I wondered if there was a pattern to these events, and began to map out locations of major unexplained events in the state. I found that there was indeed a line of activity which closely mirrored 49 Highway going North to South along the western side of the state. Then I found events located along three lines going east to west. I suddenly realized that there may be a correlation to lies of latitude and longitude and shared this idea with other researchers, including Debbie Ziegelmeyer, Chuck Zukowski, Chase Kloetske, and others.

Kansas City is at 39 degrees latitude. Somewhat coincidentally, we receive a large number of paranormal and UFO reports along 39th street in Independence, Missouri for many years. One of the most interesting cases that occurred on this street since 1947 is covered in the book "Family Secrets," by Jean Walker.

I have not been able to figure out exactly why there is so much strange activity along this street, or what the significance of the number 39 may be, if any. This line of latitude intersects with 94 degrees longitude, which correspond to two major Leylines.

In south Missouri along 37 degrees latitude, which includes the Joplin (or Hornet) Spook light, the Marley Woods, and Piedmont. These areas and other areas along 37 degrees have a higher concentration of Sasquatch and other cryptid sightings, animal mutilations, crop circles, and UFO sightings.

In mid-Missouri the 38 degrees latitude line also produces multiple effects including all of the above

mentioned anomalies. After mentioning this theory to Chuck Zukowski, he checked his animal mutilation cases in Colorado and found that they also were occurring mainly along 39, 38 and 37 degrees. Chuck did further research on 37 degrees and found that this actually extends across the United States. We also found that when Missouri had a cattle mutilation case on 39, 38, or 37 degrees latitude, a similar event occurred in Colorado on the same line within two weeks.

Why are these events occurring more often at certain lines of latitude? I needed to know more, so I Remote Viewed the situation. I got the idea while doing an RV session for another investigator on a past case where a plane was literally knocked out of the sky and crashed upside-down. For this process, I put myself into a light trance state, and send my consciousness out to look at something in the past, present or future. In this case, I saw a large craft eject a big ball of white light and it followed bluish/white grid lines from above the central U.S. clear across to California, where it hit the rear end of a commercial jet and caused it to flip over and crash.

I believe, based on my research, that UFO's may use the electro-magnetic energy grid as a form of propulsion to move about. These grid lines correspond with lines of latitude and longitude as well as some major Leylines. Ley lines are lines of energy created by underground water sources, magnetic fields, and perhaps for some reasons we don't yet understand. Where grid lines or Leys cross, they may actually create inter-dimensional portals.

The orbs of light, at least in some cases, may simply be a natural occurrence, or they could be craft of some time. This is of course, only a theory and a difficult one to prove at that due to the anomalous nature of the events that occur in these locations.

If true, the grid theory could be a real eye-opener for researchers. It could explain why certain areas, such as

Kansas City, are UFO and paranormal hot spots.

Leylines

Leylines are paths of energy that stream in different directions. They almost always run in a straight-line path. The width of the lines can vary greatly in size and strength. When major Leylines cross they are called energy centers, vortexes, portals, or places of power.

It is notable that the strongest Leyline crossing is at the pyramids in Egypt. There are several large Leyline intersections in the Unites States, with Kansas City being in one of these intersections (see next page).

In the early twentieth century Alfred Watkins discovered Leylines in England. In a vision he saw these lines run through the landscape, they ran through several places such as stone circles (Stonehenge is a major intersection), menhirs, dolmens, churches, monasteries, castles, and other buildings. In doing research he found several places with the word Ley in it, and he thought that it meant protection or lee (shelter). Because of this he called the lines "Ley lines".

Prior to this period, ancient people knew about Leylines and even dowsed for them using dowsing tools in order to find the best locations to place their buildings, or paths and roads so they could gather energy to travel.

I have checked for Leyline intersections in different locations across the state and found that many courthouses and major buildings are built on top of these locations. Perhaps the builders knew about these power centers and wanted to increase the power of the events that occurred in the buildings, just as they did in ancient times.

I've also found many Masonic symbols on these buildings. If anyone would know about ancient mysteries such as Leylines, it would likely be the Masons.

Also interesting to note is that there are two major power centers in Independence, where much of the UFO activity has been reported. These are the Truman Courthouse on the Independence Square, and the Community of Christ spiral temple. At the temple, the Leylines extend out in a circular pattern, which is highly unusual. The shape of the building is likely the cause of this as it is in a spiral shape. I've had a number of people dowse to double-check my findings and they all have the same results. Could these buildings be power

centers and possibly portals to other dimensions?

Early European woodcut of a man dowsing for Ley lines

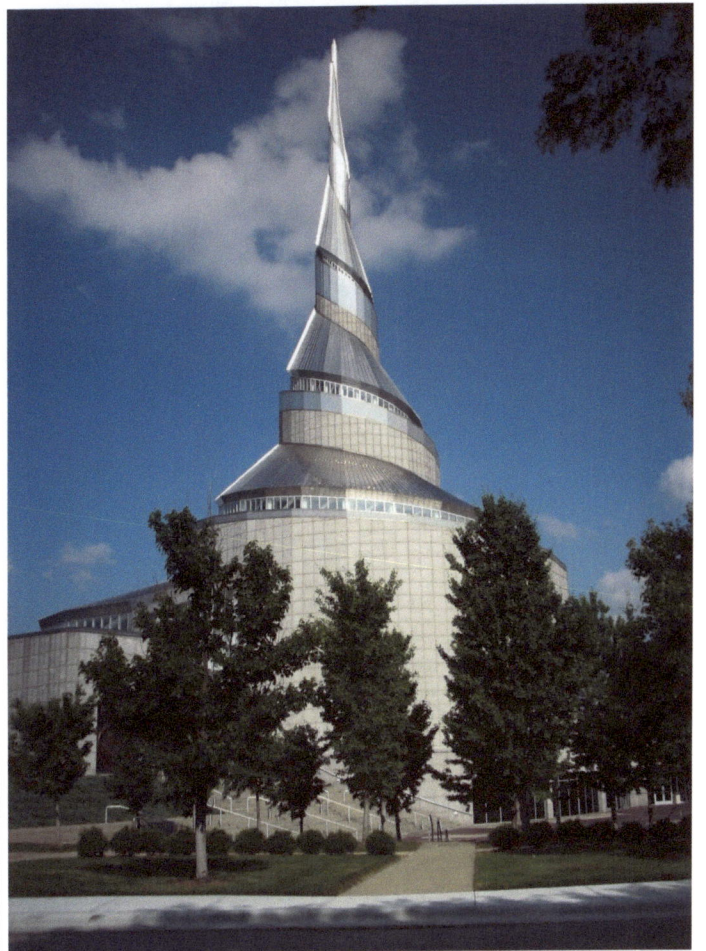

Community of Christ Temple in Independence (formerly RLDS Temple)

Photo: By Ecjmartin1 - Own work, Public Domain, https://commons.wikimedia.org/w/index.php?curid=10728539

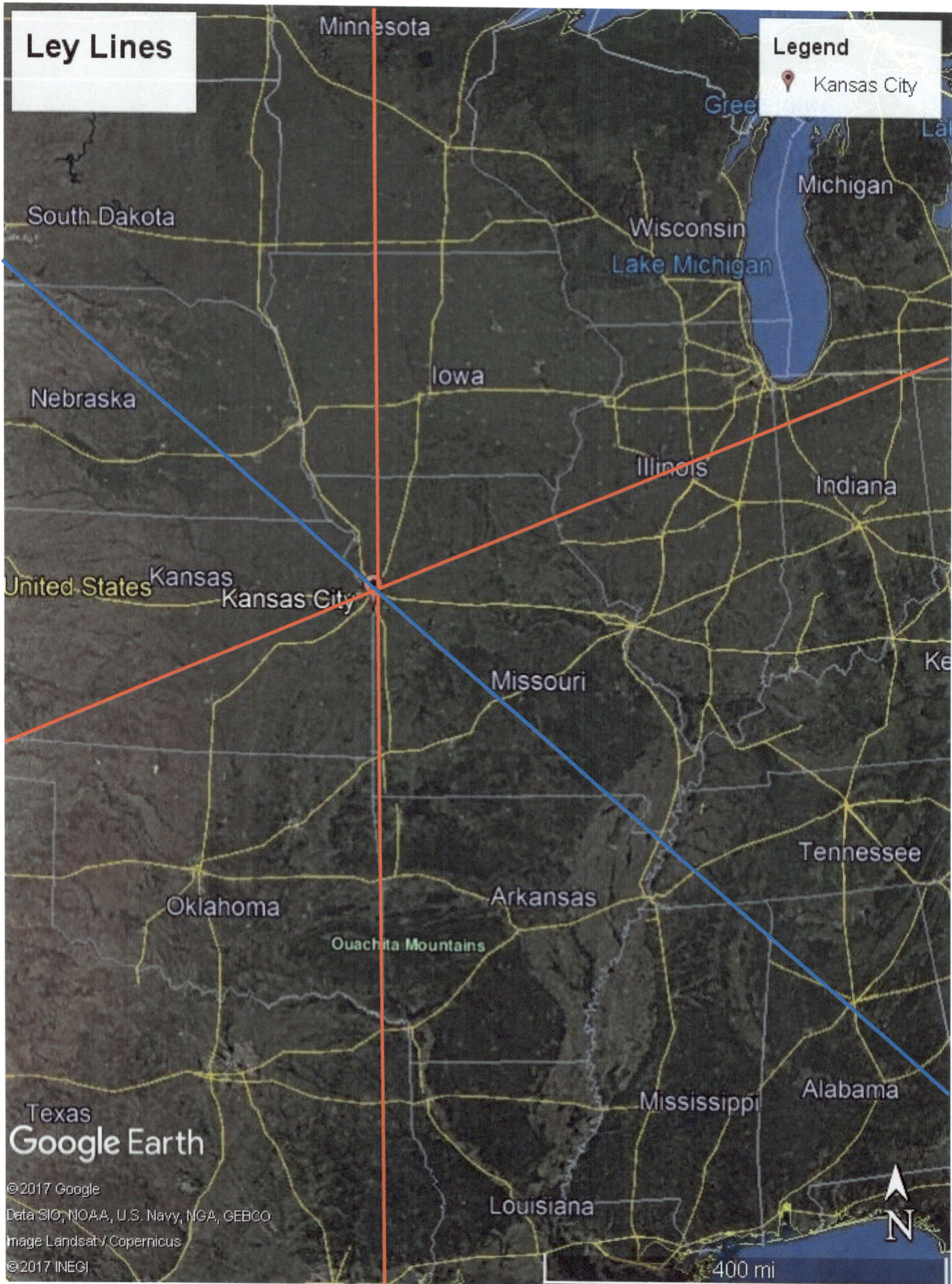

Ley Lines

Legend

📍 Kansas City

Major Leyline crossing at Kansas City. According to other researchers, there is a third major line going NE to SW as well

Latitude and Longitude

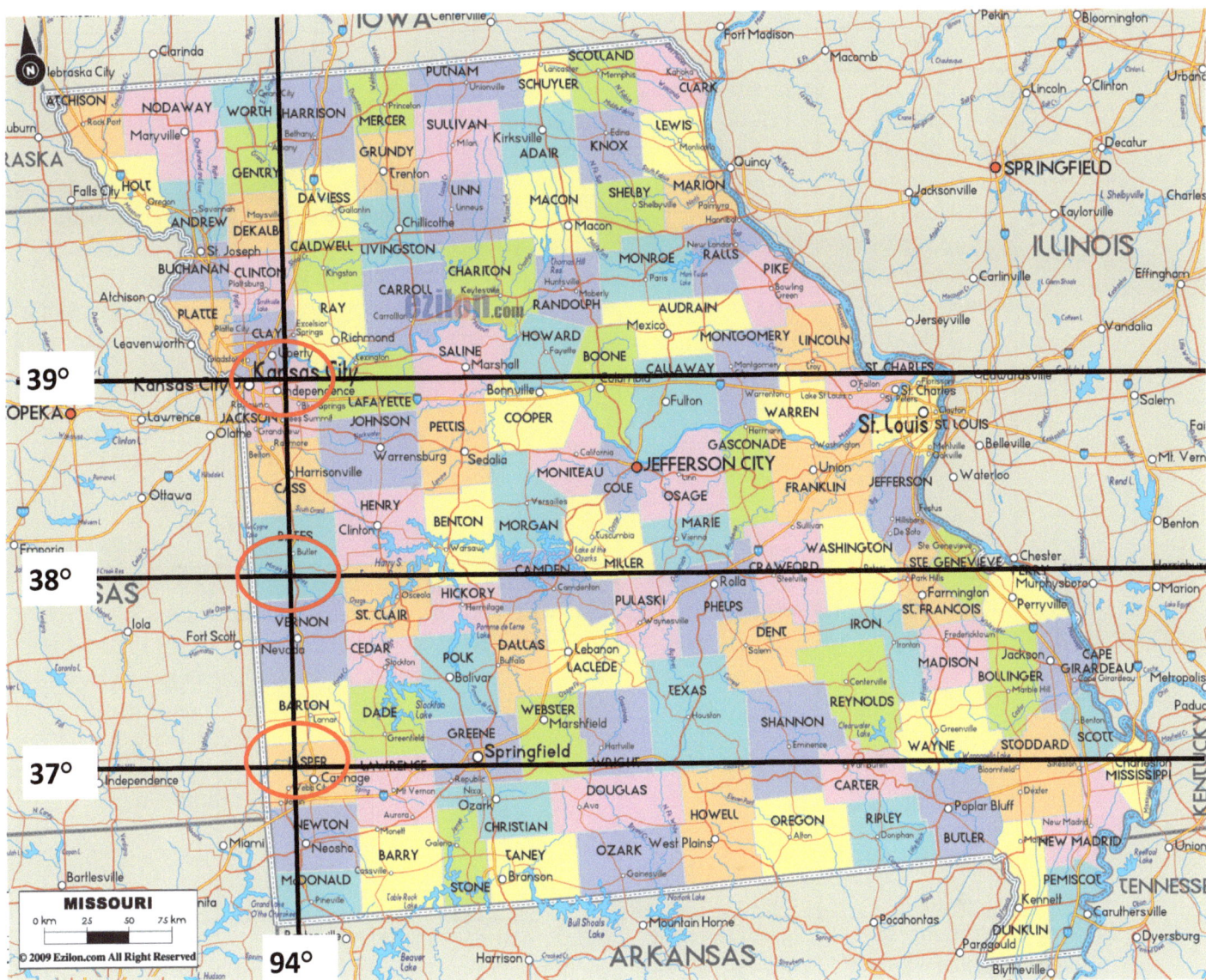

The above map roughly shows the lines of latitude and longitude, which correspond to the electro -magnetic Earth Grid. Red circles indicate UFO and paranormal hot spots on the western side of the state. Note: the remainder of the state has other hot spots as well (not shown).

UFO Coding by Jacques Vallee

Code	1	2	3	4	5
AN	AN1:Anomalies which have no lasting physical effects. i.e. amorphous lights, unexplained explosions. - Equivalent to Hynek code NL or UX, strangeness 0	AN2: Anomalies which do have lasting physical effects. i.e. poltergeists, materialized objects, areas of flattened grass, corn circles. - Equivalent to Hynek code TC, strangeness 0	AN3: Anomalies with associated entities. i.e. ghosts, yetis, spirits, elves and other mythical/legendary entities. - Equivalent to Hynek code CE3, strangeness 0	AN4: Witness interaction with the AN3 entities. i.e. near-death experiences, religious miracles and visions, out-of-body experiences. - Equivalent to Hynek code CE3 or CE4, strangeness 9	AN5: Anomalous reports of injuries and deaths. i.e. spontaneous human combustion, unexplained wounds as well as permanent healing that results from a paranormal experience. - Equivalent to Hynek code AM or CM, strangeness 9
CE	CE1: UFO comes within 500 feet of the witness, but no after effects are suffered by the witness or the surrounding area. - Equivalent to Hynek code CE1, strangeness 5	CE2: A CE1 that leaves landing traces or injuries to the witness. - Equivalent to Hynek code CE2, strangeness 5	CE3: Entities have been observed on the UFO. - Equivalent to Hynek code CE3, strangeness 7	CE4: The witness has been abducted. - Equivalent to Hynek code CE4, strangeness 9	CE5: CE4 which results in permanent psychological injuries or death. - Equivalent to Hynek code CE2 or CE3 or CE4, strangeness 9
FB	FB1: A simple sighting of a UFO traveling in a straight line across the sky. - Equivalent to Hynek code NL or DD, strangeness 2	FB2: FB1 accompanied by physical evidence. - Equivalent to Hynek code CE1 or CE2, strangeness 5	FB3: A fly-by where entities are observed on board (rare). - Equivalent to Hynek code CE3, strangeness 5	FB4: A fly-by where the witness experienced a transformation of reality into the object or its occupants.	FB5: A fly-by which the witness would suffer permanent injuries or even death.
MA	MA1: A UFO has been observed which travels in a discontinuous trajectory. i.e. vertical drops, maneuvers or loops. - Equivalent to Hynek code NL or DD, strangeness 4	MA2: MA1 plus any physical effects caused by the UFO. - Equivalent to Hynek code CE2, strangeness 5	MA3: MA1 plus any entities observed on board. i.e. the airship cases of the late nineteenth century. - Equivalent to Hynek code CE3, strangeness 5	MA4: Maneuvers accompanied by a sense of reality transformation for the observer. - Equivalent to Hynek code CE4, strangeness 9	MA5: A maneuver that results in a permanent injury or death of the witness.

Categories of Sightings

AC - Sightings From Aircraft - Pilots are considered the most experienced observers

AM - Animal Effects Cases

AN1 - Vallee Anomalies AN1 - Mysterious phenomenon

AN2 - Vallee Anomalies AN2 - Physical manifestations

AN3 - Vallee Anomalies AN3 - Fantastic beings

AN4 - Vallee Anomalies AN4 - Interacting with other beings

AN5 - Vallee Anomalies AN5 - Fatal phenomenon

BF - Bigfoot sighting

BH - Black Helicopter sighting

CE1 - Close Encounters of the First Kind - UFO's up close

CE2 - Close Encounters of the Second Kind - UFO's that leave physical effects

CE3 - Close Encounters of the Third Kind - Encounters with aliens

CE4 - Close Encounters of the Fourth Kind - Abducted by aliens

CE5 - Vallee Code CE5

CM - Cattle Mutilation

CR - Crashed Disc

DD - Daylight Discs - UFO's seen at a distance at daytime

DL - Daytime light

DO - Daytime object

FB1 - Vallee Flyby FB1

FB2 - Vallee Flyby FB2

FB3 - Vallee Flyby FB3

FB4 - Vallee Flyby FB4

FB5 - Vallee Flyby FB5

IFO - Identified Flying Object

Top UFO Websites

The following are the top UFO websites where more information is available about ufology and experiencers:

www.nidsci.org/comprehensive-catalog-1500-project-blue-book-ufo-unknowns: Project Blue Book

www.MUFON.com: Official website for the International Mutual UFO Network. List of the most recent cases from around the world, contact information for all states and countries, the MUFON radio and TV shows, annual conference and more.

www.nuforc.org: The National UFO Reporting Center is operated by Peter Davenport. A listing of recent and historical UFO reports, articles, noteworthy cases, and radio appearances by PeterDavenport is on this site. Make a report here trough the online report form or call the hotline at 206-722-3000.

www.ufohub.net: Interviews with people involved in ufology and related subjects such as aliens, space, Sasquatch, crop circles, metaphysics and more. Multiple video interviews with investigators and experiencers. Photos submitted by witnesses, and a list of events. Submit your own events here. Operated by Adnan.

www.missourimufon.org: Official website for Missouri MUFON with meeting information for St. Louis, Columbia, Kansas City, Joplin and other areas. List of UFO conferences and events, press releases and news stories. Contact information for the Kansas City ET Experiencers Group.

www.ufoevidence.org: UFO photos, drawings, cases, reports, list of UFO researchers, and multiple articles. Sighting reports are taken here.

www.ufosightingsdaily.com: Military sightings, pilot sightings, alien faces, police sightings, videos, articles, and links, live chat room, and UFO reports. Operated by Scott C. Waring.

www.ufocasebook.com: Research files, special cases, documents, photos, videos, aliens, recent sightings reports and links to other sites.

www.presidentialufo.com: The Presidents UFO Website—what the presidents know about UFOs, books, quotes, news, articles, papers, cartoons, FOIA files, and links. Register to receive updates. Operated by Grant Cameron.

www.nicap.org: National Investigations Committee on Aerial Phenomena: Global sighting information data, free online books, how to report a UFO.

www.history.com: Watch the two seasons of Hangar 1 here!

ww.theicd.com/ufo: UFO DNA– The Encyclopedia of UFO Sightings : Listing of sighting reports, categories of sightings

www.theuforeportcenter.com: Blog, multiple articles and videos, free movies, multiple subjects

www.latest-ufo-sightings.net: Sighting reports and videos posted here, including more from Kansas City

To see more graphics and information about Leylines visit:

https://www.pinterest.com/pin/225531893816718126/?lp=true

http://liminalthresholds.blogspot.com/2008/04/earth-energy-ley-lines.html

How to Obtain Evidence and Report a UFO Sighting

1. If possible, get photos or video of the object(s) in question. Ideally, us a tripod or lean against something to steady yourself. It is very difficult to analyze video tape of an object bouncing around the sky so it is important to keep the camera as steady as possible. The better the camera, the better the pictures will be. If using a cell phone keep your photo size and quality on the highest level. For video, take auto focus off and use manual focus because the auto focus will keep moving the lens in and out in an attempt to focus.

2. Make a note of the time during the sighting and after the sighting is over.

3. Sit down immediately and draw what you saw using colored pencils.

4. Write down exactly what you were doing when the sighting occurred, who was with you, and what you saw and heard in as much detail as possible. Humans forget details after some time has elapsed, and in many sighting cases don't remember that they wrote certain things down just days later. Remember that a UFO sighting can be very stressful, and the mind may block some things out.. Draw what you saw with colored pencils on white paper.

5. Make not of any missing time, weather conditions, and planes or helicopters in the area before, during, or after the sighting.

6. Do not contaminate an area that is involved in the sighting such as a location where a landing occurred, where plants were affected, or a being was sighted by walking over or near the site. This will greatly help the investigation process. In some cases people have been harmed by residual substances or radiation, so don't go near the site. Let the trained investigators do that.

7. Keep a journal after the sighting. After a sighting strange things may continue to happen so you'll want a record of that.

8. If you see a USO—Unidentified Submburged Object in water or going into our out of water, contact Debbie Ziegelmeyer, Director for the International MUFON Dive Team, MUFON Board of Directors, and State Director of Missouri MUFON. E-mail: mufondiver@aol.com

File your report immediately at www.mufon.com. You will be contacted by a local investigator by phone or e-mail. If there has been a landing or if a craft has affected trees, grass, or buildings, etc. the investigators need to get on site as soon as possible in order to collect evidence. The investigator will keep your information confidential. The investigator will ask you and all of the witnesses a series of questions for the report.

You may also want to file a report at the **National UFO Reporting Center** operated by director Peter Davenport at www.nuforc.com.

If an animal mutilation or paranormal activity is involved please report it to www.animuteinv.blogspot.com or send an email directly to me at margiekay06@yahoo.com.

Where UFO Abductees and Experiences Can Get Help

In the Greater Kansas City area:

The ET Experiencers Group meets once per month in Kansas City. This is not group therapy – just a safe place to discuss personal contacts with extraterrestrials with others who have had similar experiences. All participants are screened by the group facilitators in order to assure the confidentiality of the members. All members must live in the greater Kansas City area. All members have had encounters with UFOs and/or ET's, and most have ongoing experiences. The group is not associated with any particular UFO organization. Contact Jean Walker at jmwblondie@hotmail.com for more information.

Groups in Other Areas: Contact MUFON at www.mufon.org for a list of therapists and groups approved by MUFON. Kathleen Marden heads up this department. Also try doing a Google search for other associations or groups in your area. Sometimes just attending meetings with others who have had similar experiences can help a lot. Know this—you are not alone!

MUFON Meetings

Missouri MUFON is one of the most active UFO associations in the United States. All meetings are open to the public and are held monthly in St. Louis, Kansas City, Columbia, and Joplin, and periodically at other locations. Speakers present lectures in person or via Skype at most meetings. Some sections have annual sky watches, holiday parties, and book exchanges. The section directors present updates at each meeting so members will have current information about recent UFO sightings and encounters. Missouri MUFON sometimes collaborates with other associations to present a large conference. Visit www.missourimufon.org for more information about all of the meetings and events held across the state. Visit www.mufon.com for contact information in other states and countries.

Sources:

Un-X Paranormal Investigation Group: www.margiekay.com
UFO DNA: www.thecid.com/ufo/k.htm
The National UFO Reporting Center website: www.nuforc.org
International MUFON website: www.mufon.com
Walters, Edward UFOs are Real: Here's the Proof, Avon Books, New York, 1997

And the websites listed on page 69

Glossary of Terms

Altitude: Height above sea level

Angel: RADAR return from weather or other unknown causes

Angel Hair: Filaments found after a sighting

Angular Size: The apparent visual size (in degrees) of an object

Angular Size Scale: Estimates angular size (star, distant plane, full moon, # moons)

Angular Velocity: The apparent speed of an object in degrees/second

Anomalous Propagation: Bending of RADAR waves by the atmosphere

Area 51: One of the original plots for atomic bomb testing, later used for secret aircraft development

Azimuth: Compass bearing in degrees (0 degrees = North, 90 degrees = East)

Barium Cloud: A colored chemical cloud high in the atmosphere released by a rocket

Blue Book: Third codename for the Air Force UFO investigation project (1953-1969)

Cattle Mutilation: Dead mutilated cattle– often occurring after a UFO sighting

Crop Circle: Flattened crops or grass caused by unknown forces– often appearing after a UFO sighting

Daylight Disc: J Allen Hynek's term for a distant UFO seen in daytime

Ducting: Bending of RADAR wave by the troposphere to strike ground targets

Ecliptic: The path the planets follow through the celestial sphere

Electromagnetic Effects: Aberrations on electrical, magnetic, or radio equipment

Elevation: Angle above the horizon (0 degrees = horizon, 90 degrees = straight up)

Falling Leaf Effect: A swaying motion some UFOs make (symptom of a Fire Balloon)

Flying Saucer: A disc shaped UFO

Glitter: Sparkles that float down from the sky onto vegetation after some UFO sightings

Grudge: Second codename for the Air Force UFO investigation project (1949-1953)

IFO: Identified Flying Object

Lens Flare: Bright light reflected by lens surfaces in a camera, giving a false image

Majestic 12: An unproved report on UFOs made by a committee of 12 experts

Mogul: Project using clusters of balloons with microphones to spy on Soviet nuclear tests

Mother Ship: A UFO that releases or takes in other UFOs (ETH term)

Mirage Conditions: Atmospheric conditions that bend light, forming an image where no object is

Multiple Echo: RADAR wave bounces off 2 targets (far echo of close targets)

Nadir: The astronomical direction for straight down

Nocturnal Light: J Allen Hynek's term for a distant UFO seen at night

Non-Prosaic Explanation: Extraordinary cause -- without an ordinary explanation

Orb: A ball of light that is unexplained if seen by the naked eye. Some appear on flash photos at night and may be dust.

Orthotetny: Aimee Michel's straight line Flying Saucer theory

OVNI: UFO (Spanish" Objecto Volador No Identificado, or similar Italian, French, or Portuguese acronym)

Physical Evidence: Trace evidence left by a UFO such as glitter, a gel-like substance, plants affected, damage to the ground, trees, etc.

Prosaic Explanation: Common, mundane, ordinary cause for a UFO sighting such as the planet Venus

RADAR Sighting: Sighting only on RADAR, with no visual sighting

RADAR-V: J Allen Hynek's term for a UFO seen visually and on RADAR

Re-entry: A satellite or part of a rocket that burns as it enters the atmosphere

Rod: A rod-shaped light that may or may not be anomalous

Scout Craft: A UFO released by a Mother Ship for reconnaissance purposes

Scramble: To quickly get planes in the air to chase a UFO or other aircraft (USAF term)

Second Trip Echo: RADAR echo returns after next pulses is sent (far things look close)

Sighting: An event where someone sees a UFO

Sign: First codename for the Air Force UFO investigation project (1947-1949)

Sky Lantern: A firework with a 700 years old history that is a miniature paper hot-air balloon and often mistake for a craft

Spook Light: A Nocturnal Light that returns to the same place repeatedly

UAO: Unidentified Aerial Object

UAP: Unidentified Aerial Phenomenon

UCT: Uncorrelated Target -- RADAR target with no known air traffic there

UFO: Unidentified Flying Object -- any unrecognized object in the sky

UFO Detector: Compass with electric contacts to detect magnetic fields

UNFO: Unidentified Non-Flying Object

USAF: United States Air Force (Acronym)

USO: Unidentified Submerged Object or object seen going into or out of water

UTC: Coordinated Universal Time (Acronym) -- International (Greenwich) time

Weather Balloon: A 6 .. 10 ft. balloon released to study upper air and wind direction

Weather Blip: RADAR return caused by weather

Zenith: The astronomical direction for straight up

Afterword

When I decided to write this book I knew that I could not possibly include all 400+ investigations I've completed in greater Kansas City, so I had to narrow it down to just a few of the more significant cases in order to give the reader an idea of what is happening in this area without overwhelming him/her with details. There are many more good cases going back to the 1940's, and even further back to the 1800's that remain unexplained. Perhaps someday I'll revisit the Kansas City UFO sightings and close encounters for a future book. For now, I have two more books in the works that must be completed.

To say that the business of being a paranormal investigator is very strange, is an understatement, but it is also so interesting that I can't imagine not doing this work. There are so many things that science does not yet understand or have an explanation for. After years of working in this field I've come to the conclusion that we may never have all the answers, but we are making some strides, and with every investigation we get a little more of the big picture. Each time I investigate a case, something new comes up, but I've noticed similar patterns as well.

For instance, It is very common for people who have seen unidentified aerial vehicles to also have poltergeist type activity in their house. And often a person never has only one sighting of a UAV, they have multiple sightings, while their friends have none. Their children and grandchildren also often have similar experiences. If one sees an extraterrestrial, other family members often remember seeing the same thing or become very upset. Do the occupants of these underfitted craft follow bloodlines? Many researchers believe that they do.

My upcoming book Missouri Unexplained covers many incidents of high strangeness in the state in some of the biggest paranormal hotspots. The book includes Sasquatch sightings, hauntings, aliens, and other non-human entities, portals, and more. As most investigators will say, there is not often only one thing happening at a location, there are often many strange things that occur. And the upcoming book ExtraOrdinary is about my own experiences from childhood, which are many, and definitely fit into the high-strangeness category.

I started UnX Media Publishing company in 2016 in order to publish my own books and other's. We are seeking submissions of good non-fiction and fiction books about any unexplained subject. Please visit unxmedia.com for more information.

I am always interested in hearing about strange encounters with the paranormal. If you have an interesting story to share please send it to one of the email addresses listed on the next page.

About the Author

Margie Kay is an author, publisher, and paranormal investigator. She owns and operates a specialty chimney contracting company and forensic investigation company in Independence, Missouri along with her husband, Gene. Margie is the author of 13 books including *Gateway to the Dead: A Ghost Hunter's Field Guide*, *Haunted Independence*, *The Kansas City UFO Flap*, ExtraOrdinary, and *Missouri Unexplained*. She was the publisher and editor of Un-X News Magazine and Host of Un-X News Radio for two years. Margie is currently working on a documentary about strange events in the state of Missouri. She is the CEO of UnX Media Publishing Company.

Margie Kay is the Director of UnX Paranormal Investigation Group, which investigates haunted sites, animal mutilations, bigfoot sightings and UFO sighting reports. Kay is a psychic investigator and has worked on missing persons and unsolved crimes for law enforcement and private investigators. She has helped solve more than 50 crimes in the U.S. and abroad. Margie and the Un-X team of six have completed over 600 paranormal investigations worldwide including sites in the continental U.S., Canada, Alaska, Mexico, Chile, England, Scotland, and Ireland. Over the years, Kay has also been involved in a number of Bigfoot and animal mutilation investigations, which are sometimes related to UFO sightings.

Margie is the Assistant State Director for Missouri MUFON (Mutual UFO Network) and has completed over 450 UFO investigations to date. She is the newsletter editor and web designer for Missouri MUFON, and a Star Team Investigator for MUON.

Kay has spoken at over 100 conventions and meetings for regional and national organizations on various topics. She is a dynamic and entertaining speaker and keeps the audience engaged in the subject. To arrange for a speaking engagement or book signing please contact Margie's assistant, Jamyi Thompson at 816-833-1602 or office@unxmedia.com.

For more information visit the following websites:

Margie Kay's official website: www.margiekay.com

UnXMedia: www.unxmedia.com

Un-X News Blog: http://unxnews.blogspot.com

Animal Mutilation Investigation and Research: www.animuteinv.blogspot.com

Missouri Bigfoot Sighting Reports: http://mobigfoot.blogspot.com

On Facebook:

Kansas City Lights UFO Watch
UFO Watch Midwest
Un-X Paranormal Investigation Group
Margie Kay

UN-X MEDIA

Publications by Un-X Media:

Haunted Independence by Margie Kay 2013 – 2016
Family Secrets by Jean Walker 2017
The Kansas City UFO Flaps by Margie Kay 2017
Mysterious Missouri by Margie Kay (coming soon)
Un-X News Magazine 2011-2016 in print
Un-X News Magazine 2016-2017 digital online at www.unxnews.com
More publications in the works

Visit www.unxmedia.com for more information

Un-X Media is currently taking book submissions.
We publish non-fiction books about unexplained phenomena.
Please check the website for writer guidelines.

Contact:
editor@unxmedia.com
816-833-1602
www.unxmedia.com

Like us on Facebook at Un-X Media

UNXMEDIA

PUBLISHING